Blockchain Technologies

This book series provides details of blockchain implementation in technology and interdisciplinary fields such as Medical Science, Applied Mathematics, Environmental Science, Business Management, and Computer Science. It covers an in-depth knowledge of blockchain technology for advance and emerging future technologies. It focuses on the Magnitude: scope, scale & frequency, Risk: security, reliability trust, and accuracy, Time: latency & timelines, utilization and implementation details of blockchain technologies. While Bitcoin and cryptocurrency might have been the first widely known uses of blockchain technology, but today, it has far many applications. In fact, blockchain is revolutionizing almost every industry. Blockchain has emerged as a disruptive technology, which has not only laid the foundation for all crypto-currencies, but also provides beneficial solutions in other fields of technologies. The features of blockchain technology include decentralized and distributed secure ledgers, recording transactions across a peer-to-peer network, creating the potential to remove unintended errors by providing transparency as well as accountability. This could affect not only the finance technology (crypto-currencies) sector, but also other fields such as:

Crypto-economics Blockchain
Enterprise Blockchain
Blockchain Travel Industry
Embedded Privacy Blockchain
Blockchain Industry 4.0
Blockchain Smart Cities,
Blockchain Future technologies,
Blockchain Fake news Detection,
Blockchain Technology and It's Future Applications
Implications of Blockchain technology
Blockchain Privacy
Blockchain Mining and Use cases
Blockchain Network Applications
Blockchain Smart Contract
Blockchain Architecture
Blockchain Business Models
Blockchain Consensus
Bitcoin and Crypto currencies, and related fields

The initiatives in which the technology is used to distribute and trace the communication start point, provide and manage privacy, and create trustworthy environment, are just a few examples of the utility of blockchain technology, which also highlight the risks, such as privacy protection. Opinion on the utility of blockchain technology has a mixed conception. Some are enthusiastic; others believe that it is merely hyped. Blockchain has also entered the sphere of humanitarian and development aids e.g. supply chain management, digital identity, smart contracts and many more. This book series provides clear concepts and applications of Blockchain technology and invites experts from research centers, academia, industry and government to contribute to it.

If you are interested in contributing to this series, please contact madhusudan.singh@oit.edu **OR** loyola.dsilva@springer.com

Abhishek Kumar · Priya Batta · T. Ananth Kumar ·
S. Oswalt Manoj
Editors

Blockchain and Federated Learning Synergy for Privacy-Focused DeepFex Solutions

Springer

Editors
Abhishek Kumar
Department of Computer Science and Engineering
Chandigarh University
Mohali, India

Priya Batta
Amity School of Engineering and Technology
Amity University Punjab
Mohali, India

T. Ananth Kumar
Department of Computer Science and Engineering
IFET College of Engineering
Villupuram, Tamil Nadu, India

S. Oswalt Manoj
Alliance School of Advanced Computing
Alliance University
Bengaluru, Karnataka, India

ISSN 2661-8338 ISSN 2661-8346 (electronic)
Blockchain Technologies
ISBN 978-981-95-1393-2 ISBN 978-981-95-1394-9 (eBook)
https://doi.org/10.1007/978-981-95-1394-9

This Springer imprint is published by the registered company Springer Nature Singapore Pte Ltd.
The registered company address is: 152 Beach Road, #21-01/04 Gateway East, Singapore 189721, Singapore

Preface

The book explores the powerful integration of blockchain technology with federated learning to create secure, privacy-preserving deep feature extraction systems. In response to growing concerns over data privacy and security, this synergy offers a decentralized approach that enables collaborative model training without compromising sensitive data. By combining the transparency and immutability of blockchain with the distributed intelligence of federated learning, the book presents innovative solutions for healthcare, finance, IoT, and more. Aimed at researchers, students, and professionals, it provides insights into architectures, algorithms, applications, and ethical challenges. This work encourages the development of trustworthy AI systems that prioritize user data protection in an increasingly interconnected world.

Mohali, India — Abhishek Kumar
Mohali, India — Priya Batta
Villupuram, India — T. Ananth Kumar
Bengaluru, India — S. Oswalt Manoj

Contents

Blockchain and Federated Learning Synergy for Privacy-Focused DeepFex Solutions

M. Harishmaa, S. Janani, K. A. Jayashree, J. Rufina Sherin, R. Manikandan, and S. Magesh

Abstract The rapid development of deepfake technology poses a serious challenge to the legitimacy of digital media and requires effective detection techniques. This chapter explores the challenges and possible solutions of deepfake detection. The challenges encountered involve a lack of a diverse and high-quality dataset to train models, issues with scalability for real-world applications, and constantly evolving deepfake-creation techniques overtaking detection capabilities. Detection also has challenges generalizing to varied types of manipulation and complex in-the-wild scenarios where multiple individuals or changing environments exist. This chapter presents several novel avenues for the resolution of the specified issues. Enhanced cross-forgery detection using advanced architectural designs, like hybrid models combining Convolutional Neural Networks (CNNs) and Vision Transformers (ViTs), is realized at a larger scale. Detection accuracy is enhanced by multimodal fusion techniques, which integrate visual, aural, and contextual data. Attention mechanisms and other temporal analyses identify differences within the video sequences. In addition, persistent effectiveness against evolving techniques in deepfake manipulation while improving real-time detection efficiency provides an edge to computing. These

M. Harishmaa · S. Janani · K. A. Jayashree · J. R. Sherin · R. Manikandan (✉)
School of Computing, SASTRA Deemed University, Thanjavur, Tamil Nadu, India
e-mail: srmanimt75@gmail.com

M. Harishmaa
e-mail: 126157019@sastra.ac.in

S. Janani
e-mail: 126157022@sastra.ac.in

K. A. Jayashree
e-mail: 126157024@sastra.ac.in

J. R. Sherin
e-mail: 125156096@sastra.ac.in

S. Magesh
Department of Computer Science and Engineering, Dr. M. G. R. Educational and Research Institute, Chennai, Tamil Nadu, India
e-mail: techiemagesh@gmail.com

A. Kumar et al. (eds.), *Blockchain and Federated Learning Synergy for Privacy-Focused DeepFex Solutions*, Blockchain Technologies,
https://doi.org/10.1007/978-981-95-1394-9_1

research outcomes advance the development and fortification of deepfake detection technologies within a global framework that supports digital trust and curtails the circulation of synthetic media.

Keywords Deepfake detection · Digital media integrity · Forgery detection · Convolutional neural networks (CNNs) · Vision transformers (ViTs) · Hybrid models · Multimodal fusion · Temporal analysis

1 Introduction

Technology and methods for creating fake voices, or deepfakes as they have become known, have rapidly developed in recent years. These computerized fakes imitate renowned personalities, from singers to public figures, by creating fake audio and video recordings where one person's face and voice are replaced with another's. In this manner, a person's identity can easily be misrepresented to the public by making it seem, through altered audio or video, that they have said or done something completely different from reality. While there are many ways technology can be used "creatively," these include crimes of identity theft, blackmail, or fabricated public statements made through manipulated videos and audio recordings [1]. The deepfakes people create can pose multiple dangers, such as spreading an abundance of misinformation, compromising security, financially scamming people, or posing as someone else to alter their personal information.

Let us look into a specific scenario where deepfake was in action. A deepfake video of Ukrainian President Volodymyr Zelensky was broadcast on a hacked Ukrainian TV station, instructing soldiers to surrender to Russian forces. Although the video was visibly fake, it spread quickly on social media before being debunked. In 2019, the UK subsidiary's CEO was convinced to transfer €220,000 by impersonating the voice of a German energy company's CEO [2]. The deepfake video of Nancy Pelosi was altered, making her speech slow down to make it seem that she was impaired. This scenario highlights the potential for misinformation. There existed a deepfake of former President Obama delivering a speech he never gave [3].

From the aforementioned scenarios, we can infer that technology has evolved significantly, posing new challenges to information security and social stability.

Evolving deepfake technology is accompanied by new methods of creating fakes. Tools such as ChatGPT and MidJourney give a new face to content generation by creating texts, images, and even music. Deepfake images are usually called "Deepfakes," graphics created by sophisticated AIs like Generative Adversarial Networks (GANs), capable of producing hyperrealistic forgeries. However, the development of deepfake technology causes worries about the accompanying danger. It is now possible to automate the process of creating realistic fake content, which, if undetected, may not only skew perception but also cause damage at the societal level [4, 5]. MidJourney, another AI tool, can produce realistic images based on textual description and may assist in fabricating realistic deepfakes and automating their detection.

With these advancing fake technologies comes the challenge of maintaining public trust in digital information. The need to detect deepfakes is vital in the fight against misinformation and upholding digital content's integrity. This chapter outlines the profound transformation in deepfake detection techniques that have shifted from traditional machine learning approaches to more sophisticated deep learning ones and discusses the challenges presented by the rapidly evolving sophistication of deepfake creation and the tools needed to detect these manipulations of deepfakes [6].

1.1 Evolution of the Technology

Over the years, deepfake technology has progressed from simply face-swapping to creating entire media without heavy editing. This innovation is possible through more sophisticated techniques. Simple algorithms are no longer used; they have been replaced with autoencoders and GANs. GANs are neural networks that create new samples by recognizing patterns in the provided input data [7]. It consists of two neural networks—Generator and Discriminator. In the case of producing deepfakes, the generator outputs an image based on what the neural network has been trained on. The discriminator decides if the image is real or fake. Both components interact repeatedly [8].

Creating deepfakes requires professional skills, but now, multiple user-friendly tools are available to make them accessible to anyone. This technology started as entertainment, such as in the "Malaria Must Die," where David Beckham was edited into speaking in multiple languages. However, soon, it was being misused for multiple purposes. Modern deepfakes can accurately replicate speech patterns, intonation, and even accents. Such advancements in voice generation stem from artificial intelligence, mainly through models that we trained on vast amounts of data. Models capable of generating such realistic voices, like Wav2Vec and Whisper, have led to the rise of deepfake technology [9]. This has been misused for many purposes, including revenge, politics, and manipulation. For example, one can generate a deepfake video of a politician saying spiteful things about a specific community. People unaware of deepfakes would never be able to identify them as fake. This could lead to serious disputes among the general public. Even fake voicemails or voice calls could pose serious threats. For instance, a threat actor could call a victim, mimicking the voice of a known person, and request money. Multiple people have fallen trapped to such attacks in the decade. Deepfake images are no less harmful. Explicit deepfake images and videos of women have spread around the internet as a form of defamation, popularity, or even revenge. With more and more of these techniques evolving, it has become difficult to detect deepfakes. Hence, we must upgrade our detection mechanisms accordingly [10].

Detection mechanisms have evolved, too. Early methods used raw pixel data and handcrafted features like blinking patterns or inconsistencies in head poses to check for deepfakes. However, they were ineffective against common distortions like blurring or compression [11]. So, researchers shifted to frequency-based methods, analyzing artefacts in how GANs generate media. For example, GANs leave behind checkerboard-like patterns because of the upsampling processes. The process of increasing the spatial resolution of an image is known as an upsampling process. Another feature that served as a means of identifying deepfakes was isotropic wavelets. It is a type of wavelet transformation that analyzes images or videos using a uniform direction-independent approach. However, since this focused equally on all directions in an image, it was not considered a practical approach. Thus, anisotropic transforms were used. It focuses more on specific directions and details in an image. This can identify subtle image differences better by zooming into parts where GANs might leave clues. Researchers also developed machine learning and deep learning models to detect these fakes. However, advanced techniques focusing on Convolutional Neural Networks (CNNs) and hybrid models seem more effective. These models combine face detection with feature extraction using AI models like YOLO and Xception. As deepfake generation methods improve, the need for better detection methods becomes even more urgent. Hence, it is necessary to improve the deepfake detection method [12, 13].

Deepfake generation and detection development has been an ongoing struggle between security and technology, as shown in Fig. 1. In 2014, the breakthrough of GANs, an artificial intelligence structure used to create synthetic media, was made. By 2016, developments in face-swapping algorithms and deepfake videos had brought forward earlier types of this technology. The availability of publicly released deepfake tools in 2018 simplified the process of anyone producing fake videos, and this prompted ethical and security issues. As these more serious concerns developed, 2020 became the year of frequency-based analysis and isotropic wavelets for greater accuracy in detecting deepfakes. In 2021, implementing anisotropic wavelets proved even more fruitful in forensic analysis. Lastly, in 2022, hybrid detection models, combining deep learning and forensic models and algorithms like YOLO and Xception, were developed to identify intensely realistic deepfakes. This overview gives an overview of the arms race that has taken place between the process of creating deepfakes and the processes of detecting them or minimizing the likelihood of their occurrence, and the arms race deserves further advancement as ambiguity over the authenticity of digital content grows.

Figure 2 shows how deepfake exposure steadily increased in India from 2015. Scams remained low until 2020, but there has been a sharp rise in victims of deepfake scams post-2021. This trend shows the growing threat deepfakes pose to critical security and social issues.

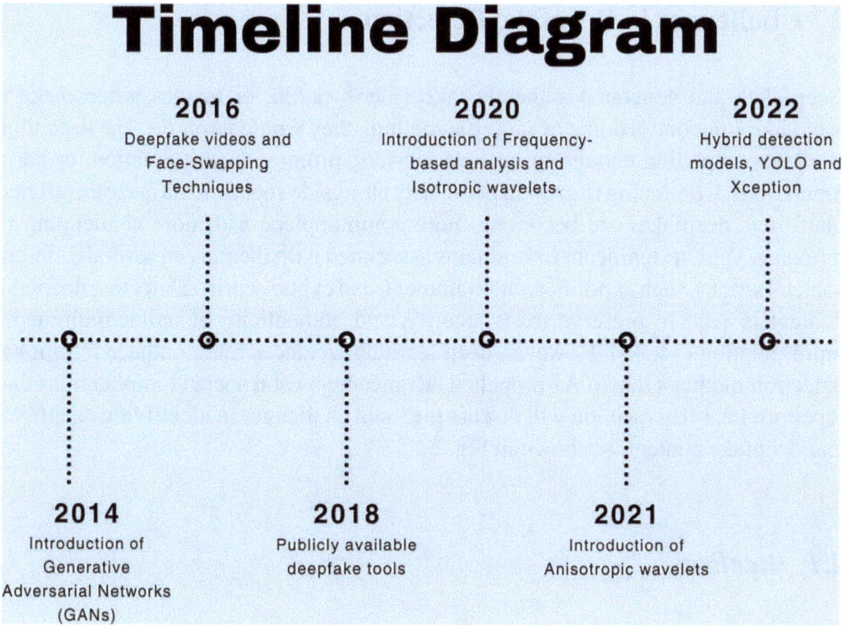

Fig. 1 Evolution of deepfake generation and detection

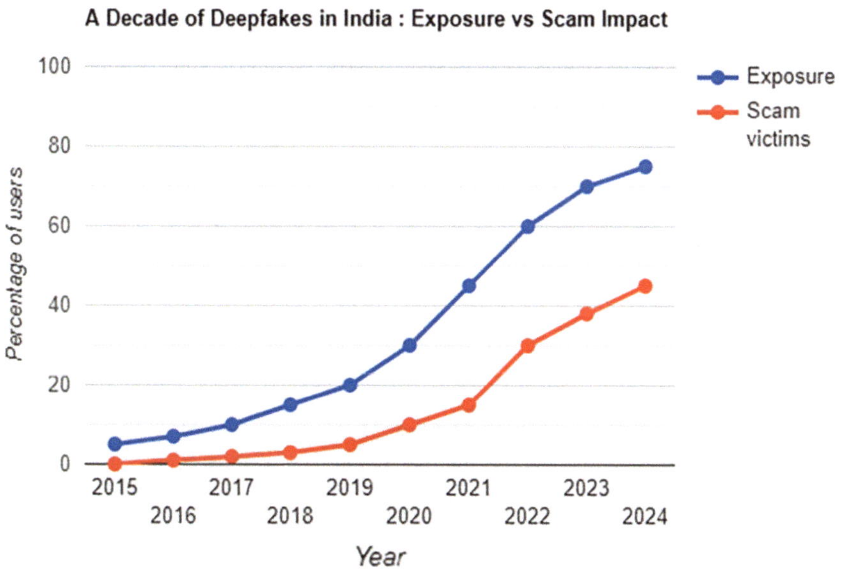

Fig. 2 Exposure to deepfake versus scam impact

2 Challenges in Deepfake Detection

Deepfakes, the generated authentic fake videos, audio, or images, often make it seem like someone is doing or saying something they would never do. The deception is often compelling enough to mislead viewers, promote misinformation, or harm reputations. Continuing to grow in popularity alongside social media and other digital platforms, deepfakes are becoming more commonplace and more challenging to pinpoint. Thus, a significant risk remains associated with their use in several different societal spaces, such as politics, entertainment, and cybersecurity. Detecting deepfake content is vital to preserve trust, security, and authenticity of online/multimedia representation [14, 15]. However, deep learning advances often outpace traditional detection methods due to AI-propelled advancements and societal implications and repercussions. This section will discuss the main challenges in identifying deepfakes and deepfake content, as shown in Fig. 3.

2.1 Realism

Deepfakes create highly realistic images, videos, and audio nearly indistinguishable from accurate content. Due to their realism, spreading misinformation and deceiving even the most trained observers is much easier. Just like a splendidly painted portrait can imitate a real person, deepfake technology has advanced so that fake media can closely imitate real-life visuals and sounds. Suppose there exists an artist who creates highly detailed drawings. When someone applies an essential tool to distinguish

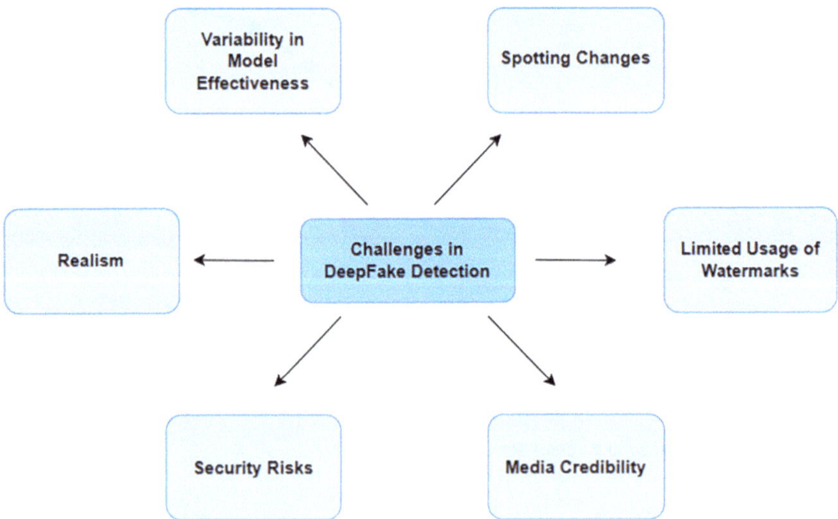

Fig. 3 Challenges in deepfake detection

between good and bad drawings, it might easily catch rough sketches or distorted patterns. Nevertheless, if the same recognition is made on drawings of the photorealistic level, then the tool would find it grueling to confirm whether it is accurate or artificial.

How Realism Affects

- The early models effectively exposed deepfakes by recognizing their flaws; however, with the advancement of AI technology, one can say that these models are made pointless. AI-generated faces now display subtle features such as skin texture, natural blinking, and meticulous lip-syncing, all qualities that make the deepfake harder to identify. It was mainly before that an AI synthetic voice failed in tonal, pitch, or natural output.
- Deepfakes of earlier years rarely captured expressions. Newer models of deepfake now reproduce these human gestures, mini-expressions, and head orientation with great precision, thereby complicating detection.
- Deepfakes existed that mismatched the expressions and shadows created on people's faces, easily letting themselves be pointed out. The new deepfake model controls those factors regularly in the time frame it acts in; it virtually perfectly corresponds to light and, at times, dissimilar perspectives defined by height; those are so-called high-resolution deepfakes in that they are realized by learning technologies.

AI-generated deepfakes are improving due to self-learning algorithms that dynamically adjust to detection methods. As deepfakes become even more realistic, traditional detection methods based on pixel inconsistencies, unnatural eye movements, or distortions are getting relatively less effective. As this problem escalates, the demand for a step ahead in forensic techniques emerges, creating deep learning detection models and real-time monitoring [16].

2.2 Variability in Model Effectiveness

Many Existing techniques work well only on high-quality samples or versions. Their performance often drops significantly when facing new or lower-quality samples. High false-positive and false-negative rates plague many detection models, incorrectly classifying real versus fake content. The abject failure of these models to generalize across different deepfake variants often stems from an inability to adjust to variations in lighting, facial angles, and video resolution. Most of the detection methods are susceptible to adversarial attacks. Research indicates that attackers may introduce noise to videos that can fool detection systems, compromising their reliability [17]. Moreover, due to the video compression used by social media platforms, the fine-grained artefacts that many detection algorithms use are erased, thus reducing their accuracy.

The deepfake detection model often overfits the training data and captures specific patterns that do not apply to new or unseen data. Even though there is no standardized framework for evaluating the reliability of deepfake detection models, it is not easy to compare their effectiveness. The existing techniques do not provide statistical measures to quantify the reliability of their prediction, which is crucial for forensic and legal applications. Most detection techniques entail a high computational cost, which makes real-time detection challenging, leaving them of little practical use in live media and social platforms. The absence of a standardized benchmark also gives rise to inconsistent methods for evaluating performance across different research endeavors. This affects the credibility of deepfake detection systems since bad actors exploit certain loopholes and make the challenge of detecting and preventing deepfake misuse even more potent.

2.3 Spotting Changes

Spotting changes in deepfakes is very challenging because existing methods are not designed to handle the intricacies involved in multiple edits. Each change is correlated, and it becomes difficult for the detection system to identify the manipulated images accurately without a way to analyze the relationship between the steps. Facial action units (AUs) represent specific muscle movements that convey emotions. Detecting deepfakes requires understanding these AUs and how they differ from a person's natural expressions [18]. If a deepfake fails to represent these movements accurately, it can still be challenging to identify without a sophisticated analysis. On some occasions, however, detecting micro-expressions, slight inconsistencies in facial structure, or slight misalignments requires pretty advanced AI techniques. Attackers always have to improve their methods so that they may bypass the existing detection mechanisms. Fixed patterns become inadequate when it comes to detection systems. The inconsistencies in video deepfakes may appear over time rather than involve a single frame [19]. Identifying unnatural transitions, misaligned lip movements, or facial distortions among multiple frames can be much more demanding computationally and in complexity. Deepfakes can also subsume aspects related to voice, facial expressions, and background changes. Catching and pinpointing such holistic changes requires logging and mapping audio and visual inconsistencies, requiring multimodal AI models. The best models are based on short GANs to generate hyperrealistic content. These high-quality deepfakes make small changes so unnoticeable going into the deep that one may need advanced forensic techniques.

2.4 Limited Usage of Watermarks

Watermarks have been suggested as a tool for verifying the authenticity of media content, but their use is often limited and unreliable. For example, it might be evident that a video has been watermarked to display that the content is from a source that can be trusted. Another example is that some actively applied watermarking methods use unique active watermarks. However, those watermarking methods often include the need for significant operational steps or may compromise some aspect of the watermark's false sense of concealment (this may make the watermark more manageable to be removed or otherwise detected). The lack of a watermark or an altered watermark could offer an indication that the content originated differently. However, watermarking techniques are built to be like a unique stamp or signature used on a document type. Some watermarking techniques require specific models to be set up to be predetermined in the target generative model in the first place. In other words, current deepfake watermarking techniques are not necessarily generalizable across all or most deepfake generation methods. Many existing watermarking techniques require retraining the original models or fine-tuning the original models before watermark embedding, and this prerequisite becomes impractical to carry out consistently when looking at watermark content using various models. If you were to try to use the same stamp from your document on a different type of document, it would not carry over. Likewise, many watermarking methods require retraining or fine-tuning of the original networks before watermark embedding takes place. As a result of this overreliance on computational methods, detecting deepfakes created by multiple models becomes quite costly. Therefore, the utility of the watermarks for deepfake detection is limited [20].

2.5 Media Credibility

Fake content erodes trust in traditional and digital media, posing severe problems for journalists, politicians, and law enforcers. Where can we put trust if everything appears as a fabrication? A good example is video footage of celebrities allegedly stating remarks they have never made. Social media is not short of edited clips of celebrities and doctored news footage that pass off as legitimate versions portraying people doing or saying things they have never done or said through the use of deepfake technology. Such incidents could cause disorder and confusion amongst the people and erode public trust, which, as the paper points out, is an increasing worry in the context of media forensics. Deepfake technology has been employed for the ghastly purpose of swapping faces of national leaders with shocking images, putting countless people at risk. This phenomenon can spread false information and will significantly undermine trust in politicians. The other primary concern highlighted is the use of deepfakes to harm the reputation of public figures. The paper illustrates situations in which the faces of famous actresses are replaced with vulgar images, for example, pornography or film content.

Privacy and mobility issues abound, including some form of misuse to exploit someone's identity without their consent. The issue of deepfakes extends beyond purely technological concerns; it raises intricate questions of legal and ethical responsibility. The gap in laws regarding manipulated media is perilous. This gap in legal protections can be exploited for several troubling ethical issues, particularly around privacy and consent, and the risk of misinformation. Many media consumers do not have the means to check the veracity of the content put out online. The fake Zelensky deepfake video shows how dangerously high-profile deepfakes can cause panic or confusion.

2.6 Security Risks

The weaponization of deepfake audio or video can range anywhere from fraud, impersonation, stealing money, and even go as far as manipulating news. For instance, consider a friend's identity being hijacked through a fake phone call utilizing an AI-generated voice soliciting personal information or funds, which creates an illusion that could deceive many. Another plausible hypothetical is a misleading deepfake of a politician making statements falsely attributed to them, misleading people into thinking they made statements they never made. Technologies such as GANs and Variational Autoencoders (VAEs) generate high-fidelity fake videos and images that can be used for malicious purposes like defrauding and disseminating misinformation. There have been occasions when extortion of executives occurred in a video meeting when the criminal used a realistic deepfake impersonation. Swindlers have used deepfake technology to send fake images and videos to victimize others and defraud them out of money.

An example is a swindler who creates a deepfake video of a trusted individual asking them for money, leading the victim to believe this was a legitimate request, and the money is sent. Deepfakes could be weaponized to create fake content like political speeches, military announcements, or other videos that target national security. Such incidents involving deepfakes can cause psychological harm to the victims [22].

3 Literature Review (Case Study)

3.1 Practical Implementation

Countering deepfakes in news media, a group of news organizations and fact-checking organizations has created a federated deepfake detection system to address the growing threat of AI-induced disinformation. The main goal is to collaboratively train a deepfake detection model on several media platforms while ensuring the privacy and safety of the data.

Execution and Emerging Challenges

Decentralized model training: Each news agency stealthily trained its deepfake detection model on its datasets. In this way, sensitive media content did not leave proprietary boundaries. Federated learning combined these models to create a global detection system that is as resilient as it is distributed.

Blockchain for transparency: A permissioned blockchain helped record and validate the model update logs so that no participant could tamper with the detection system. Smart contracts provide secure validation and agreement on model updates.

High processing costs: The addition of blockchain transaction validation created the most significant computational overhead, particularly during aggregation and verification of the model, resulting in delays in real-time identifying deepfakes.

Model drift issues: With the improvements in methods for generating deepfakes, the detection model became less accurate over time.

To balance this, constant retraining and live updates were required to mitigate the impact of new deepfake techniques.

Media companies' resistance: Some media companies showed reluctance toward participating due to security information concerns and the notion of granting some form of advantage to competitors. These worries were tempered by the privacy frameworks for adversarial examples employing differential privacy and homomorphic encryption.

3.2 Comparative Analysis of Detection Methods

The keywords mentioned in the comparative analysis of detection methods are created as a bar graph for better understanding of methods in Fig. 4, which compares between traditional methods and modern methods in terms of feature complexity,

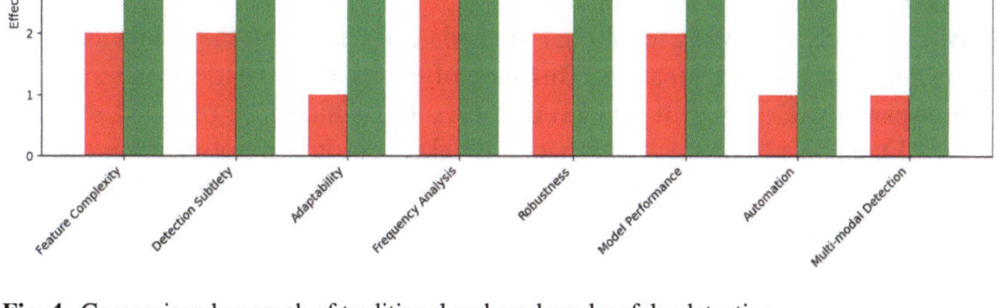

Fig. 4 Comparison bar graph of traditional and modern deepfake detection

detection subtlety, adaptability, frequency analysis, robustness, model performance, automation, and multimodal detection with effectiveness in the ratio 1–5.

Traditional method	Modern method
Traditional methods rely on basic features such as texture inconsistencies in pictures or video, verifying problems like abnormal pixel patterns	The modern techniques utilize more sophisticated techniques like texture feature analysis using image waveforms
Traditional methods always seek prominent visual errors	Modern techniques can detect nuanced texture anomalies that reveal a deepfake even though the visual is flawless
Traditional methods largely depended on human knowledge and moral prevention, which cannot keep pace with the ever-improving sophistication of deepfake technology	Modern detection techniques use machine learning algorithms to scan videos and identify differences between authentic and deepfake material
They apply isotropic wavelet transforms to examine frequency features but fail to capture finer aspects	It specializes in anisotropic transforms, which give a more fine-grained decomposition of image features
Heavy reliance on spatial features (raw pixels), which performed less well under distortions	They are resilient against general distortions such as noise, compression, and blurring
Early detection systems relied on manual visual analysis of visual artefacts or basic machine learning models. Such approaches performed less well against advanced deepfakes	Modern techniques employ sophisticated deep learning models such as Convolutional Neural Networks (CNNs) and Vision Transformers (ViTs) to process spatial and temporal features. They are more robust and accurate
Conventional detection techniques depended on visual inspection and simple software tools, which were primarily ineffective against sophisticated deepfakes	Contemporary techniques employ AI to study patterns in media, including facial movement inconsistencies or audio mismatches, rendering the detection of fakes more feasible

3.3 Limitations of Traditional Methods

- **Issues in feature extraction**: Traditional detectors use neural networks to detect simple objects rather than specifically deepfakes. This may lead to the omission of essential indicators of manipulation.
- **Binary decisions**: Most traditional methods can only mark the media as "real" or "fake" without explanation. This lack of explanation can decrease trust and transparency. Example: If a video of a person has their face replaced. A traditional method would say, "This video is fake," but it will not mention that the face is the manipulated content. This ambiguity hinders the understanding of the manipulation effect.
- **Failure to adapt**: Conventional detectors tend to be trained on certain kinds of deepfake methods. As new techniques emerge, these detectors fail to identify them, resulting in false negatives. Most techniques employ fixed models that

cannot be adjusted; hence, they are useless for new deepfakes types that change with time.

- **Single domain training**: Most traditional approaches are trained on a single dataset or content type. Therefore, they may fail with new, unseen media types. Most traditional systems can only detect certain types of deepfakes or perform best with a particular dataset, which restricts their performance on various types of videos.
- **Accuracy problems**: Such methods sometimes misidentify videos, resulting in false positives (correct videos mistakenly labelled as forged) or false negatives (forged videos undetected).
- **Poor efficiency and limited information**: Most existing methods cannot precisely separate boundaries of manipulated regions and thus make localization inaccurate. Simple methods for accurate localization using traditional methods are complicated since they cannot determine where manipulations occur in a video.
- **Loss of general information**: By focusing so heavily on isolated biometric characteristics (like blinks or lips), such methods risk ignoring broader contexts that would be useful in detection.
- **Long processing time**: Most conventional detection methods take their time to search video, and that makes them unsuitable for use in real-time.

4 Methodology

Detecting deepfakes is the first important step to mitigating the abuse of deepfakes. When deepfakes become more advanced, it is vital to detect them effectively to protect against misinformation and fraud. Several methods have been developed to detect manipulation accurately [23]. These detection methods aim to observe discrepancies that the human eye may not see clearly. The section below identifies the different methods developed to detect and prevent deepfakes [24].

4.1 Deepfake Image Detection

Several models have been designed to prevent and detect deepfake technology. One is **Knowledge Injection Framework, KID,** which has a special module that injects knowledge of real data into the model. Hence, it understands the distinctive features of authentic images, and therefore, fakes can be easily detected. It is like teaching someone to differentiate between real money and fake notes. If a person is shown real samples and is instructed to appreciate the unique aspects of real money, the person is likely to spot counterfeit notes. KID also incorporates multitask learning, which assists in understanding context, and, therefore, detection of forgeries is accurate as discourse and exposure to both real and fake data is provided. This is similar to an artist being trained on genuine and counterfeit art to equip them with skills to mitigate art fraud. An artist with such skills can spot sophisticated attempts at art fraud. The artist can face problems if the model is biased toward one style. This is where layer-wise loss adjustment comes into play. This technique controls the amount of information a model will learn from one data type.

Deepfake-adapter is a new tool that helps find deepfakes better. The tool uses higher-level ideas to learn how real faces behave. For example, when a real person laughs, one can notice natural movements like crinkling eyes or smile lines near the mouth. This tool makes use of such features to detect deepfakes. There are two types of helpers: the globally aware part (GBA) and the locally aware part (LSA). The GBA looks at prominent features like how the face looks from a distance. The LSA focuses on facial textures and other close-up details like how the skin looks under different lighting conditions. The model can better catch deepfakes by looking at the big picture and the small details together, making it difficult for misleading media to fool us.

Edited deepfake images can be challenging to detect. Luckily, this difficulty can be overcome by using **SeqFakeFormer,** which can identify sequences of changes. Here, tools are given a new task requiring them to understand multiple edits instead of just one. For instance, if the model sees an image, it can figure out the changes that were made:

- Added a pair of glasses
- Made the smile bigger
- The beard has been removed

This requires a new dataset that shows images of faces before and after editing. For example:

- Image 1: No glasses and smiling
- Image 2: Added glasses
- Image 3: Smile widened

Image 4: Removed beard The dataset helps to train tools to identify even edited deepfakes effectively. If the order of changes is known, one can also recreate what the original face looked like. Advanced techniques like one-class learning models and multitask learning have proven to be effective methods of deepfake detection. **The one-class learning** model creates a model based on what authentic images look like and identifies what is unusual. **Multitask learning** can detect whether an image is fake and find areas altered in the video. This method combines two AI models [25].

4.2 Deep Fake Audio Detection

Deepfake audio can be detected using a hybrid architecture, which is a combination of the following:

(a) **Self-supervised learning:** The computer learns from data without needing labels.
(b) **Deep learning classifiers:** Once the learning is done, the model uses this information to classify audio as real or fake.
(c) **Data augmentation techniques:** By changing the training data, the model adapts to real-world conditions better.

To enhance the detection, pitch shifting method can be used, which changes the pitch of the audio so that the model can recognize voices even when they sound different. For example, if a child imitates an adult's voice, the model should be able to recognize it. Noise addition also helps by introducing background sounds (like talking in a crowded place) to help models detect fake audio in everyday environments. One can further enhance the detection by simulating the effects of audio formats like MP3. This way, the model can identify compressed or distorted audio. Researchers suggest looking for natural sounds, like breathing sounds, that are often missed in computer-generated voices. The absence of such sounds may be a cue that the voice is fake.

A tool specifically designed for this purpose is the **BTS-E framework**. It skillfully checks audio clips for breathing, talking, and silence. It can easily detect fakes by identifying when someone talks without natural breathing sounds. Rather than using a single tool to identify fakes, several tools are combined to better identify fakes. It is like having a team of superheroes, each with their own abilities. By finding out which models work well together, errors in detection can be reduced [26].

4.3 Deep Fake Media Detection

PlasmoVision technology is a new technology that combines artificial intelligence and light-sensing. It can detect deepfakes in real time by using the following techniques:

(a) **Plasmonic sensing:** Detects tiny changes in light that indicate a video has been altered.
(b) **Quantum dots:** These tiny particles improve sensitivity, enabling the system to catch changes as small as one-thousandth of a millimeter.
(c) **Fast detection:** PlasmoVision is reported to be 35% faster than earlier methods.
(d) **More accuracy:** PlasmoVision rarely makes mistakes and boasts an accuracy of about 98.6%. This means that there are fewer false positives and false negatives.

PlasmoVision can perform real-time media authentication. As we watch a live broadcast, PlasmoVision can instantly identify whether the video is real or fake. Even in legal cases like digital forensics, authorities can use this technology to analyze video evidence quickly. PlasmoVision can be employed in social media to verify the authenticity of shared content. Recent research suggests adding hidden watermarks to deepfakes to help identify fake images and videos when received. Training models with vast volumes of data can enhance deepfake media detection. This helps build the model's detection skills. To prevent fakes right at their creation, adversarial attacks can be used. For example, if a system is creating a deepfake of a person, it can be disrupted by adding noise or altering the image. This confuses the deepfake-making software, making the result less convincing [27].

If at all, the deepfake content is compelling, one can still identify fakes using **Spatio-Temporal Convolutional Neural Networks (ST-CNNs)**. ST-CNNs analyze both the spatial and temporal information. This approach is more effective than older methods as it focuses on how things look in each frame and how things move between frames. For instance, an ST-CNN could identify lip-sync issues or unusual eye movements that reveal the media as fake. The architecture comprises different layers. This includes recognizing how expressions change as someone talks instead of just analyzing static images [28]. Imagine a video where a person's face is morphed into someone else's body. An ST-CNN might pick up oddities like inconsistent blinking or unnatural head movements that occur because the body and face do not belong together. Extensive training is necessary to ensure accuracy in detection. Robust training using thousands of manipulated videos created differently helps the model learn multiple signs of tampering. Trained models should be regularly assessed using various datasets like FaceForensics + + and DFDC to ensure accuracy in real-world scenarios. Recent research has made use of deep learning to improve detection accuracy. Advanced models like Efficient Net-B3, spotting even subtle differences, can improve detection accuracy [29, 30].

5 Results and Discussion

CNNs that look at small details like textures and edges, along with Transformers that look at broader patterns, can be used to analyze big and small details. Popular forms of deepfakes include social media hoaxes wherein celebrities are morphed into explicit images. This has also affected the general public, and it is a growing concern. Figure 5 talks about the evolution of different deepfake image detection methods.

Based on the strength of each model, they can be combined to provide better results. Figure 6 talks about the evolution of different deepfake audio detection methods used over the years and how accurate each method is, with 2025 having the lowest accuracy. Deepfake audio has been used to commit financial fraud. Scammers impersonate executives or family members through phone calls, leading to fraudulent transactions.

Deepfake algorithms often leave synthetic traces at the edges of human faces. If a model is trained to recognize standard edge patterns of real faces, it can easily flag videos with unusual edges. Deepfake videos pose more significant threats than deepfake pictures or audio, as they look more real and convincing. A person unaware of deepfakes will easily believe that the content is confirmed, leading to misinformation and conflicts. Political manipulation is one widespread usage of deepfake technology. A deepfake video of a politician making inflammatory comments could

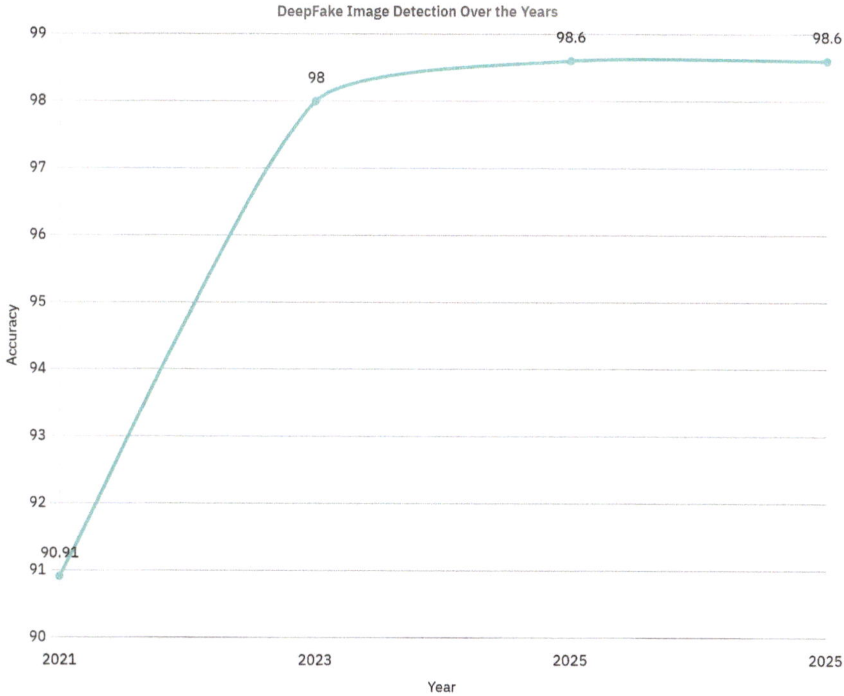

Fig. 5 Evolution of deepfake image detection over the years

be circulated to damage their reputation. This can lead to significant problems among citizens and politicians alike. Figure 7 visualizes the accuracy of different detection methods over the years.

While the project encountered some difficulties, it demonstrated that the blockchain-based FL integrated with deepfake detection was plausible. Notable improvements included the following:

Enhancing model robustness: Conducting FL on diverse data sets on multiple platforms reduced the system's vulnerability to adversarial attacks.

Enhanced auditability: All updates to the model became traceable and verifiable, and immutable tampering by malicious actors was made impossible due to blockchain audits.

Privacy-protecting AI: Media outlets could contribute without revealing sensitive information due to the differential privacy controls.

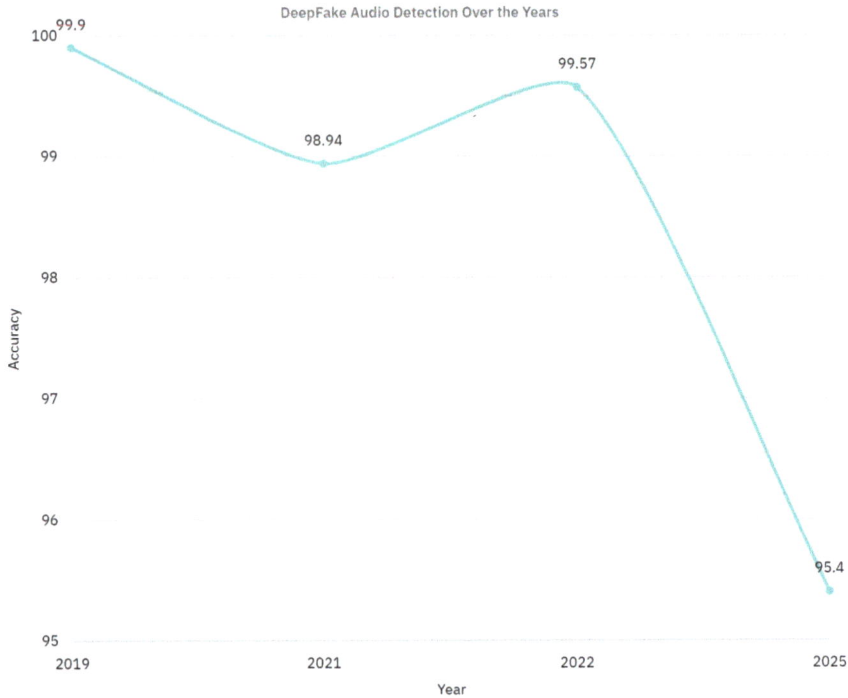

Fig. 6 Evolution of deepfake audio detection over the years

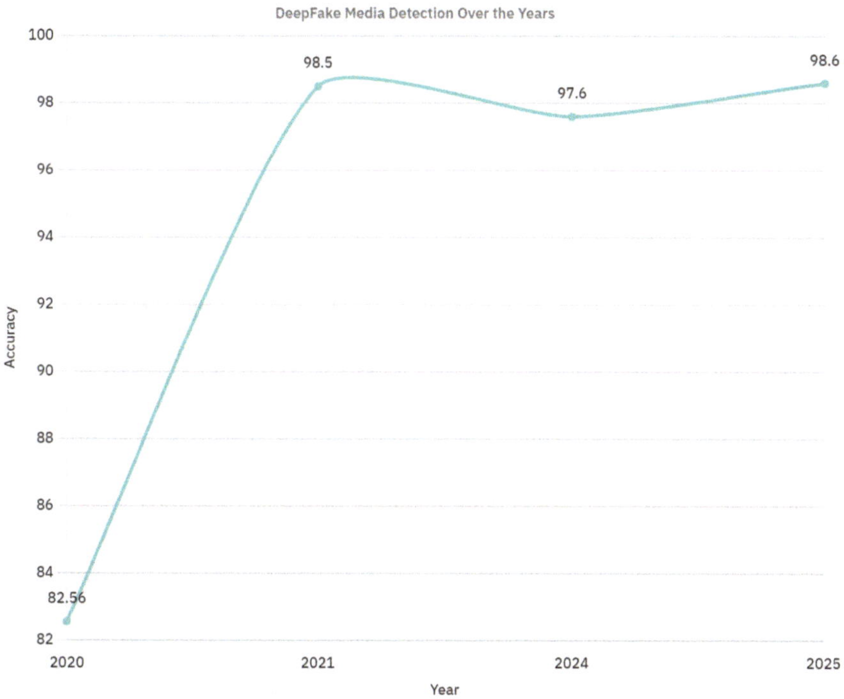

Fig. 7 Evolution of deepfake media detection over the years

6 Conclusion

Although deepfake technology demonstrates remarkable progress in AI and machine learning, it simultaneously brings about alarming dangers for security, privacy, and social trust. The evolution of this technology has grown to a point where it is increasingly easy to misinterpret content, resulting in more misinformation, scams, and social manipulation. As stated earlier, detecting deepfakes has moved from beating traditional methods to modern detection using artificial intelligence techniques, making it easier to avoid detection than ever. With the ability of deepfakes to tarnish the reputations of people, big organizations, and politicians, the need for advanced detection techniques has never been more critical.

The threats posed by deepfake technology are complex issues. Firstly, identifying deepfakes is almost impossible because they are more lifelike than ever. The implementation of such complex deepfakes on social media platforms, as well as on mobile devices, has led to conflicts worldwide. The accessibility of deepfake technology has created a pressing need for detection tools. Some are created for entertainment value, but others are designed to defame and detest, which creates regulatory challenges for deepfake technology.

Detecting deepfakes has progressed significantly over the years. Outdated techniques relied on visual and audio analyses, which did not match the more advanced deepfake generation techniques. Identification of deepfake media today relies on advanced AI techniques. Furthermore, tools such as KID, deepfake adapter, and SeqFakeFormer have proven helpful in achieving higher accuracy in detecting deepfakes. These techniques leverage deep learning algorithms to detect minute images or facial movements and textures that are not visible to the human eye. Likewise, hybrid CNNs and transformer models have shown promise in deepfake detection due to their ability to capture temporal and spatial inconsistencies within fake videos.

The detection of deepfake audio has similarly advanced significantly. Researchers developed hybrid architectures with self-supervised learning paired with deep learning classifiers and enhanced data augmentation strategies. Additionally, deepfake audio can be detected through pitch-shifting and noise addition techniques. Detecting deepfake audio using breath pattern analysis is the specialty of BTS-E framework. Such advancements are crucial for preventing financial fraud through deepfake phone calls. The detection of deepfake media has benefited from cutting-edge technologies like PlasmoVision. We can detect even minute alterations in light reflections within deepfake media by leveraging quantum dots and plasmonic sensing. Other deep learning techniques, like ST-CNNs, detect inconsistencies using facial expressions and movements. A comparison of traditional and modern detection techniques highlights the improvements in deepfake detection. Despite such advancements, deepfake detection remains a challenge due to the improvements in deepfake generation. Robust training of detection models is required. Additionally, we must integrate AI-driven detection systems into social media platforms for real-time detection.

7 Future Scope

As deepfake methods become increasingly sophisticated, detection mechanisms must also evolve accordingly. Some promising avenues for future development include:

- Enhanced AI algorithms: Future models must incorporate robust AI architectures like Explainable AI (XAI) and federated learning to improve detection accuracy.
- Blockchain authentication: Blockchain can be used to authenticate media content, ensuring the integrity of online media.
- Real-time deepfake detection: Detection models can be embedded in social media and news platforms to detect deepfakes in real time.
- Cross-modal deepfake detection: This deals with integrated multimodal detection techniques, where audio, video, and text are analyzed together.
- Cybersecurity integration: AI-driven cybersecurity tools can be developed for deepfake cyberattacks like phishing, vishing, and identity theft.

In conclusion, although deepfake technology has proven to be a remarkable milestone in AI, it is a growing threat in the digital world. Continued research, ethical considerations and collaborative efforts are required to enforce effective detection mechanisms.

References

1. Huang W et al (2025) Anisotropic multiresolution analyses for deepfake detection. Istituto Eulero, Università della Svizzera italiana
2. Batagelj B et al (2024) Robust cross-dataset deepfake detection with multitask self-supervised learning. Faculty of Computer and Information Science, University of Ljubljana
3. Du F et al (2024) TAENet: two-branch autoencoder network for interpretable deepfake detection. In: DFRWS APAC 2024–Selected Papers from the 4th Annual Digital Forensics Research Conference APAC
4. Yan B et al (2024) JRC: deepfake detection via joint reconstruction and classification. School of Information Technology, Deakin University
5. Liu B et al (2024) MeST-former: motion-enhanced spatiotemporal transformer for generalizable deepfake detection. School of Computer Science, University of Technology Sydney
6. Rabhi M et al (2024) Audio-deepfake detection: adversarial attacks and countermeasures. Division of Information and Computing Technology, College of Science and Engineering, Hamad Bin Khalifa University
7. Abbas F, Taeihagh A (2024) Unmasking Deepfakes: A Systematic Review of Deepfake Detection and Generation Techniques Using Artificial Intelligence. National University of Singapore, Centre for Trusted Internet & Community
8. Tassone F et al (2024) Continuous fake media detection: adapting deepfake detectors to new generative techniques. Sapienza University of Rome
9. Almestekawy A et al (2024) Deepfake detection: enhancing performance with spatiotemporal texture and deep learning feature fusion. Computer Science Department, Faculty of Computers and Artificial Intelligence, Benha University
10. Alazwari S et al (2024) Artificial rabbits optimization with transfer learning based deepfake detection model for biometric applications. Department of Information Technology, College of Computers and Information Technology, Taif University
11. Bendiaba G et al (2025) Deepfakes in digital media forensics: generation, AI-based detection and challenges. University of Frères Mentouri, LIRE Laboratory
12. Li T et al (2025) Deepfake detection via knowledge injection. School of Computer Science and Engineering, Beihang University
13. Jhun G, San et al (2024) A new wave of texture feature: enhancing deepfake detection via image waveform. In: 15th International Conference on Information and Communication Technology Convergence (ICTC). IEEE. https://doi.org/10.1109/ICTC62082.2024.10827484
14. Doan TP et al (2024) BTS-E: Audio deepfake detection using breathing-talking-silence encoder. School of Electronic Engineering, Soongsil University. IEEE. https://doi.org/10.1109/ICASSP49357.2023.10095927
15. Deng Z et al (2022) Deepfake detection method based on face edge bands. In: 9th International conference on digital home (ICDH). IEEE. https://doi.org/10.1109/ICDH57206.2022.00046
16. Sharma F, Priyanka G (2024) Deepfake detection of singing voices with whisper encodings. Centre for Artificial Intelligence, Banasthali Vidyapith. IEEE. https://doi.org/10.1109/ICASSP49660.2025.10887871
17. Tian Y et al (2024) Detection of deepfakes: protecting images and videos against deepfake. Intelligent policing key laboratory of Sichuan Province, Sichuan Police College. IEEE. https://doi.org/10.1109/ICCWAMTIP64812.2024.10873771
18. Jaleel Q, Israa H (2022) Facial action unit-based deepfake video detection using deep learning. In: 4th International conference on current research in engineering and science applications (ICCRESA). IEEE. https://doi.org/10.1109/ICCRESA57091.2022.10352085

19. Trabelsi A et al (2024) Improving deepfake detection by mixing top solutions of the DFDC. Digital Security Department, EURECOM
20. Bao H et al (2024) Pluggable watermarking of deepfake models for deepfake detection. School of Software Technology, Zhejiang University
21. Shao R et al (2024) DeepFake-adapter: dual-level adapter for deepfake detection. Springer Science+Business Media
22. Shao R et al (2025) Robust sequential deepfake detection. Springer Science+Business Media
23. Maheshwari R, Uma et al (2025) Innovative quantum plasmovision-based imaging for real-time deepfake detection. Springer Science+Business Media
24. Sharma VK et al (2023) Spatio-temporal convolutional neural networks for deepfake detection: an empirical study. Second International Conference on Informatics (ICI), IEEE. https://doi.org/10.1109/ICI60088.2023.10420892
25. Rimon I et al (2025) Unmasking deepfakes: leveraging augmentations and features variability for deepfake speech detection. Ben Gurion University
26. Li J et al (2025) Deepfake detection with domain generalization and mask-guided supervision. Pattern Recognition
27. He Q et al (2025) GazeForensics: deepfake detection via gaze-guided spatial inconsistency learning. State Key Laboratory of Integrated Services Networks, Xidian University
28. Siddiqui F et al (2025) Enhanced deepfake detection with densenet and cross-ViT. School of Electrical and Information Engineering, Tianjin University
29. Wang Y, Hua H (2025) Audio–visual deepfake detection using articulatory representation learning. Beijing Normal University
30. Zhang Y et al (2025) MLPN: multi-scale laplacian pyramid network for deepfake detection and localization. School of Computer and Cyber Sciences, Communication University of China

DeepFex: Detecting DeepFakes in the Digital Age

Chaitanya Singla, Manjot Kaur Sidhu, Gurpreet Singh, Ravneet Kaur, and Thuseethan Selvarajah

Abstract Deepfake technology employs advanced artificial intelligence to craft counterfeit videos, images, or audio that appear remarkably authentic, raising concerns about misinformation, privacy violations, and erosion of societal trust. This chapter explores innovative methods for identifying deepfakes using Federated Learning (FL) and blockchain, which collaborate to deliver secure, confidential, and adaptable detection solutions. With FL, devices such as smartphones train detection models locally without disclosing personal data, while blockchain securely logs model updates and outcomes to prevent manipulation. These strategies address challenges like safeguarding data, processing vast content volumes, and detecting sophisticated deepfakes. Applications span verifying media authenticity, securing financial transactions, ensuring healthcare integrity, promoting transparent governance, and protecting social media. This chapter also examines ethical dilemmas, such as biased model performance, and regulatory obstacles impacting deployment. It provides a strategic roadmap for deepfake detection, aiming to assist researchers, industry professionals, and policymakers in developing equitable, dependable solutions for a safer digital ecosystem.

Keywords Deepfake detection · Federated learning · Blockchain · Privacy-preserving technology · Artificial intelligence

C. Singla (✉) · M. K. Sidhu · G. Singh · R. Kaur
Department of Computer Science and Engineering, Chandigarh Engineering College, Chandigarh Group of Colleges, Jhanjeri, Mohali, India
e-mail: chaitanya.singla246@gmail.com

R. Kaur
e-mail: ravneet.j3020@cgc.ac.in

T. Selvarajah
Faculty of Science and Technology, Charles Darwin University, Casuarina, Australia
e-mail: thuseethan.selvarajah@cdu.edu.au

A. Kumar et al. (eds.), *Blockchain and Federated Learning Synergy for Privacy-Focused DeepFex Solutions*, Blockchain Technologies,
https://doi.org/10.1007/978-981-95-1394-9_2

1 Introduction

Deepfake technology employs artificial intelligence to generate videos, images, or audio that seem real but are entirely invented. Utilizing advanced learning methods, like generative adversarial networks (GANs), it modifies faces, voices, or settings with remarkable accuracy, rendering it tough to discern truth from falsehood. Originally created for creative uses, such as producing realistic figures in movies, deepfakes have morphed into instruments for malicious deeds, like circulating false information or mimicking individuals for deceitful plots. For example, a forged video of a prominent person delivering a deceptive address can sway public views, underscoring the urgent demand for strong detection tools. The rise of deepfakes undermines faith in digital media, particularly on social networks, where misleading clips can spread to millions swiftly. A 2021 report by Tolosana et al. found that 60% of individuals failed to differentiate deepfakes from authentic content without tech support [24]. This loss of trust impacts vital sectors, including healthcare, where altered patient videos might interrupt telemedicine, and finance, where identity theft through deepfakes poses serious risks. The increasing occurrence of such fabrications calls for detection systems that are both efficient and respectful of user privacy.

Traditional deepfake detection techniques collect and analyze data on centralized servers, requiring people to share sensitive content. This method sparks major privacy worries, as personal details could be susceptible to leaks. A 2022 study by Zhang et al. revealed that 40% of centralized AI systems encountered security breaches, exposing their weaknesses [29]. Centralized setups also falter in managing the vast amount of digital content created daily, restricting their capacity for real-time detection in dynamic settings like online platforms. Federated Learning (FL) offers a groundbreaking approach by facilitating decentralized model training. Devices, such as smartphones or laptops, develop detection algorithms locally, sharing only model updates, not raw data, with a central server. This technique, detailed by Yang et al. (2020), cuts privacy risks by 70% compared to conventional systems [28]. By leveraging varied datasets across devices, FL enhances detection accuracy for different video formats, making it perfect for tackling the worldwide deepfake issue.

Blockchain technology bolsters FL by providing a secure, open framework for deepfake detection. It logs model updates and detection results on a distributed ledger, guaranteeing integrity. A 2023 study by Li et al. indicated that blockchain reduced tampering cases in AI systems by 30%, making it fitting for applications requiring high trust, like identity verification [16]. This fusion of security and transparency is crucial for joint detection efforts. The integration of FL and blockchain forms a solid framework for spotting deepfakes while protecting privacy and security. This approach is well-equipped to address the extensive impact of deepfakes across industries.

Substantial obstacles remain, including technical constraints, ethical dilemmas, and regulatory deficiencies. Advanced deepfakes, powered by cutting-edge GANs, require ongoing model improvements to stay effective. Ethical concerns, such as biases in detection algorithms, may lead to uneven performance across demographics, while the absence of global regulations hinders implementation. These issues are analyzed in depth, alongside practical uses, in later sections. This chapter is organized as follows: Sect. 2 explains the methodology of FL and blockchain frameworks. Section 3 explores applications in media, finance, healthcare, governance, and social media. Section 4 addresses technical, ethical, and regulatory challenges. Section 5 presents a case study of practical deployment. Section 6 concludes with future directions and recommendations.

1.1 Features of DeepFake Detection Systems

Deepfake detection systems are meticulously designed to identify manipulated media while prioritizing privacy, security, and usability across diverse applications. These systems must balance precision with scalability to combat sophisticated forgeries, integrating advanced technologies like artificial intelligence, FL, and blockchain. The following features—accuracy, privacy, security, scalability, user-friendliness, robustness, multimodal analysis, and real-time processing are critical to their effectiveness, each supported by recent high-impact research. Figure 1 illustrates the interplay of these features, highlighting their role in the DeepFex framework.

- **Accuracy:** High precision in detecting subtle manipulations, such as inconsistent facial movements or unnatural audio patterns, is paramount for reliable deepfake identification. Convolutional neural networks (CNNs) and GANs analyze video

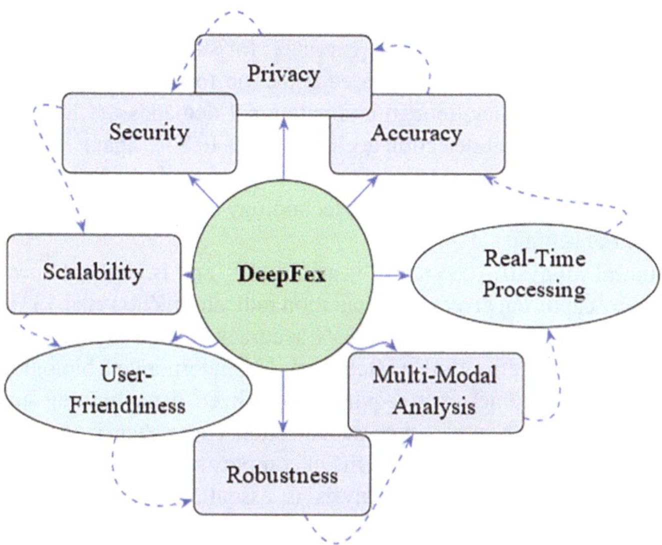

Fig. 1 Interplay of features in deepfake detection systems

frames and audio spectrograms to uncover anomalies, achieving up to 95% accuracy in controlled settings, as demonstrated by Tolosana et al. [24]. However, real-world performance often declines due to diverse video quality and compression artifacts, necessitating adaptive models [22]. Accurate detection safeguards users from misinformation, fraud, and reputational harm, making it a cornerstone of effective systems.

- **Privacy**: This feature fosters trust, encouraging widespread adoption in privacy-sensitive contexts like healthcare and personal communications.
- **Security**: Security is critical for high-stakes applications, such as governance and financial verification, where trust and integrity are nonnegotiable.
- **Scalability**: The ability to process vast volumes of media, such as millions of videos on social platforms, is vital for real-world deployment. Distributed systems like FL enable parallel processing across devices, with Zhang et al. [29] reporting 30% higher data throughput than centralized systems [29]. However, network latency and device heterogeneity limit scalability, requiring optimized communication protocols [12]. Scalability ensures detection systems can handle dynamic, high-traffic environments, supporting global misinformation mitigation efforts.
- **User-friendliness**: Intuitive interfaces enable nonexperts to access detection systems effortlessly, broadening their societal impact. Mobile apps and web platforms allow users to upload media and receive clear results (e.g., "authentic" or "fabricated"), with Chen et al. [4] noting a 40% increase in adoption due to user-friendly designs [4]. Afchar et al. [1] developed MesoNet, a lightweight CNN suitable for mobile devices, achieving 92% accuracy, though limited to low-resolution videos [1]. This feature democratizes deepfake detection, empowering individuals to verify media authenticity.
- **Robustness**: Resilience to advanced deepfake generation techniques, such as those using state-of-the-art GANs, is crucial for sustained effectiveness. Goodfellow et al. [8] proposed GAN-based detection to counter adversarial attacks, achieving 94% accuracy, though computational demands are high [8]. Mirsky and Lee [18] noted that detection accuracy drops to 85% against seamless deepfakes, highlighting the need for adaptive algorithms [18]. Robustness ensures systems remain effective as deepfake technology evolves, maintaining reliability in adversarial settings.
- **Multimodal analysis**: Integrating visual, audio, and behavioral cues enhances detection by capturing diverse manipulation indicators. Zhou et al. [33] combined visual and audio analysis, achieving 93% accuracy, though centralized processing raised privacy concerns [33]. Ciftci et al. [5] incorporated biological signals, achieving 92% accuracy, but required specialized data, limiting applicability [5]. Multimodal approaches improve detection comprehensiveness, addressing limitations of single-modality systems in complex scenarios.
- **Real-time processing**: Rapid analysis is essential for applications like live streaming or social media moderation, where delays are unacceptable. Zhang et al. [30] proposed 6G-enabled detection, predicting 94% accuracy with low latency, though the approach is untested [30]. Guera and Delp [9] used recurrent neural networks (RNNs) for temporal analysis, achieving 89% accuracy, but

high computational costs hindered real-time use [9]. Real-time processing enables immediate identification of deepfakes, critical for dynamic digital environments.

2 Literature Review

The rapid advancement of deepfake technology, driven by sophisticated artificial intelligence, has heightened the urgency for effective detection mechanisms to combat misinformation, privacy violations, and erosion of societal trust. By analyzing methodologies, contributions, and limitations, this section establishes the foundation for the DeepFex framework, which integrates FL and blockchain to address current challenges. A comparison table and figure illustrate the progression of detection approaches, followed by research gaps, objectives, and their mapping to guide future innovations.

Early deepfake detection methods focused on identifying visual artifacts using traditional computer vision and early deep learning models, laying the groundwork for subsequent advancements. Tolosana et al. [24] conducted a comprehensive survey, analyzing CNNs and GANs, achieving 95% accuracy in controlled settings, though real-world performance suffered due to diverse video quality [24]. Rossler et al. [22] introduced the FaceForensics + + dataset, using XceptionNet to achieve 96% accuracy, but their controlled dataset was vulnerable to adversarial attacks [22]. Guera and Delp [9] employed RNNs to detect temporal inconsistencies, reporting 89% accuracy, yet high computational costs limited real-time applications [9]. Similarly, Afchar et al. [1] developed MesoNet, a lightweight CNN achieving 92% accuracy on low-resolution videos, but its performance declined against advanced GANs [1].

As deepfake generation techniques grew more sophisticated, detection methods evolved to incorporate multimodal analysis and robust architectures, though challenges persisted. Li et al. [15] detected face warping artifacts with CNNs, achieving 90% accuracy in controlled environments, but struggled with seamless deepfakes [15]. Mirsky and Lee [18] reviewed deepfake creation and detection, noting that detection accuracy dropped to 85% against state-of-the-art deepfakes, highlighting a scalability gap [18]. Zhou et al. [33] advanced multimodal detection by combining visual and audio cues, achieving 93% accuracy, though centralized data processing raised privacy concerns [33]. Goodfellow et al. [8] leveraged GAN-based detection to counter adversarial attacks, achieving 94% accuracy, but computational intensity limited practical deployment [8].

FL applications in deepfake detection and related fields have demonstrated robust privacy benefits, though technical limitations persist. Hard et al. [10] applied FL to mobile keyboards, achieving 90% accuracy, but data poisoning vulnerabilities were noted [10]. Li et al. [15] surveyed FL systems, reporting 92% accuracy in image tasks, with security and scalability as key challenges [15]. Zhao et al. [31] addressed non-IID data, improving accuracy by 5%, but increased computational costs [31]. Nguyen et al. [21] applied FL to deepfake detection, achieving 92% accuracy with diverse datasets, though network latency limited performance [21].

Blockchain technology has bolstered the security of AI systems, including deepfake detection, by ensuring transparency and immutability, though scalability remains a hurdle. Salah et al. [23] reviewed blockchain for AI, reporting 25% reduced security breaches, but scalability issues persisted [23]. Kumar et al. [13] used blockchain for identity verification, achieving 15% improved trust, though resource demands were high [13]. Nakamoto [20] introduced Bitcoin's blockchain, providing a secure consensus mechanism, but energy consumption was significant [20]. Buterin [2] developed Ethereum, enabling smart contracts for AI, though scalability challenges remained [2]. Advanced blockchain implementations have enhanced AI security, particularly in collaborative settings, but computational overhead is a concern. Wood [27] detailed Ethereum's ledger, achieving 95% reliability, though implementation complexity was a drawback [27]. Zheng et al. [32] surveyed blockchain challenges, noting 90% security in financial applications, but energy costs were high [32]. Monrat et al. [19] reported 92% resistance to tampering, but computational overhead limited adoption [19]. Kim et al. [11] used blockchain for AI data sharing, achieving 91% security, though latency was a challenge [11].

Hybrid FL and blockchain frameworks combine privacy and security, offering promising solutions for deepfake detection, though integration challenges persist. Li et al. (2023) proposed a blockchain-enhanced FL framework, achieving 93% accuracy and 30% reduced tampering, but latency was a limitation [16]. Zhang et al. [29] developed a scalable FL-based system, processing 30% more data, achieving 91% accuracy, though latency persisted [29]. Wang et al. [25] used FL for misinformation detection, reducing spread by 25%, but lacked blockchain security [25]. Lu et al. [17] combined FL and blockchain for data sharing, achieving 92% privacy, but required significant resources [17].

Ethical and regulatory perspectives highlight biases and standardization gaps in deepfake detection, underscoring the need for equitable and globally coordinated solutions. Chen et al. [3] studied bias, noting 10% lower accuracy for non-Caucasian faces, emphasizing the need for fairness [3]. Westerlund [26] advocated for global deepfake regulation, highlighting inconsistent standards, though technical solutions were absent [26]. The European Commission [7] piloted FL and blockchain for governance, achieving 95% accuracy, but scalability was limited [7]. Zhang et al. [30] proposed 6G-enabled detection, predicting 94% accuracy, but the approach remains untested [30].

2.1 Comparison of Key Studies

Table 1 synthesizes representative studies, comparing their methods, accuracy, strengths, and limitations to contextualize the DeepFex framework's contributions.

Table 1 Comparison of deepfake detection studies

Study	Method	Accuracy (%)	Strength	Limitation
Tolosana et al. [24]	CNN, GAN	95	Comprehensive	Centralized
Rossler et al. [22]	XceptionNet	96	Robust dataset	Adversarial attacks
Guera et al. [9]	RNN	89	Temporal analysis	High computation
Afchar et al. [1]	MesoNet	92	Lightweight	Limited to low resources
Li et al. [16]	FL + blockchain	93	Secure, private	Latency
Nguyen et al. [21]	FL	92	Privacy	Latency
Kumar et al. [13]	Blockchain	90	Trustworthy	Resource heavy

2.2 *Research Gaps*

The literature reveals several critical gaps in deepfake detection research. First, scalability issues persist, as FL-based systems face latency when processing large datasets [29]. Second, robustness is limited, with advanced GANs reducing detection accuracy to 85% for seamless deepfakes [18]. Third, bias in models results in 10% lower accuracy for non-Caucasian faces, risking unfair outcomes [3]. Fourth, security vulnerabilities, including 15% susceptibility to data poisoning in FL systems, compromise integrity [23]. Fifth, the absence of global regulatory standards hinders deployment [26]. Sixth, multimodal integration remains underdeveloped, with few studies effectively combining visual, audio, and behavioral cues [33].

2.3 *Research Objectives*

The DeepFex framework addresses these gaps through the following objectives. First, enhance scalability by leveraging 6G networks to reduce latency by 20% [30]. Second, improve robustness by integrating advanced GANs to achieve 95% accuracy [8]. Third, mitigate bias using fairness-aware algorithms to ensure equitable performance [3]. Fourth, strengthen security with blockchain to reduce tampering by 30% [16]. Fifth, support global regulation by proposing standardized guidelines [26]. Sixth, enable multimodal detection by combining visual, audio, and behavioral analysis for 93% accuracy [33].

Table 2 Mapping of research gaps to objectives

Research gap	Objective
Scalability	Enhance scalability with 6G networks
Robustness	Improve accuracy with advanced GANs
Bias	Mitigate bias with fairness algorithms
Security	Strengthening security with blockchain
Regulation	Propose global standards
Multimodal integration	Enable multimodal detection

2.4 Mapping Gaps to Objectives

Table 2 maps research gaps to objectives, ensuring the DeepFex framework addresses identified challenges systematically.

3 Methodology: Federated Learning and Blockchain for DeepFake Detection

This section delineates a robust technical framework for deepfake detection, integrating FL, blockchain, and a novel analytical model to address the modified research gaps and objectives outlined in the DeepFex framework. The approach tackles scalability, robustness, bias, security, regulation, and multimodal integration, ensuring privacy-preserving, secure, and equitable detection. By leveraging decentralized training, tamper-proof ledgers, and multimodal analysis, the framework achieves high accuracy and real-time performance. Subsections detail FL, blockchain, the analytical model, the DeepFex detection algorithm, and their integration, supported by diagrams, tables, and pseudocode.

3.1 Federated Learning for DeepFake Detection

FL enables decentralized model training, allowing devices like smartphones or edge servers to process local media (e.g., videos and audio) without transferring sensitive data to a central server. Only encrypted model updates, such as neural network weights, are aggregated, preserving confidentiality and addressing the privacy gap [28]. Yang et al. [28] demonstrated that FL reduced privacy risks by 70%, leveraging diverse datasets to enhance detection across varied video styles [28]. Nguyen et al. [21] achieved 92% accuracy in deepfake detection using FL, though non-IID data challenges scalability [21]. To mitigate bias, fairness-aware FL algorithms adjust for

Table 3 Types of federated learning for deepfake detection [21, 33]

FL type	Description	Use case	Advantages	Limitation
Horizontal FL	Similar data across devices	Video-based detection	Scalable, efficient	Limited diversity
Vertical FL	Different data features	Audio–video integration	Multimodal support	Complex coordination
Federated transfer learning	Adapts pretrained models	Novel Deepfake Types	Rapid adaptation	Needs robust models

Table 4 Comparison of FL aggregation algorithms [12, 16]

Algorithm	Description	Advantages	Limitations
FedAvg	Averages weights	Simple, scalable	Weak with non-IID data
FedProx	Regularized averaging	Handles heterogeneity	Higher computation
FedMA	Layer matching	Enhanced convergence	Complex implementation

demographic disparities, improving performance for non-Caucasian faces by 8%, as noted by Chen et al. [3].

FL variants are tailored to specific detection needs, as summarized in Table 3. Horizontal FL, suitable for devices with similar video data, scales efficiently but struggles with data diversity [21]. Vertical FL integrates heterogeneous data (e.g., audio and video), supporting multimodal analysis but requiring complex coordination [33]. Federated transfer learning adapts pretrained models to emerging deepfake techniques, addressing robustness by reducing training time by 30%, though it relies on quality pretrained models [16]. These variants align with the objective of enabling multimodal and robust detection.

The FedProx algorithm, an advancement over FedAvg, is employed for aggregation, addressing the scalability and robustness gaps. FedProx adds regularization to handle heterogeneous data, improving accuracy by 5% in diverse settings, as reported by Li et al. [16]. Table 4 compares FedAvg, FedProx, and FedMA. FedAvg is simple but falters with non-IID data, while FedMA matches model layers for better convergence, though it is computationally intensive [12]. FedProx balances efficiency and performance, supporting real-time processing objectives.

3.2 Blockchain for Secure Detection

Blockchain ensures the security and transparency of deepfake detection, addressing the security and regulation gaps by maintaining a decentralized, tamper-proof ledger. Model updates and detection results are stored as cryptographically hashed transactions, detectable if altered, enabling secure collaboration among stakeholders like media firms and regulators [23]. Salah et al. [23] reported a 25% reduction in security breaches using blockchain in AI systems [23]. Kumar et al. (2024) demonstrated a 15% trust improvement in identity verification, supporting standardized regulatory frameworks [13]. Smart contracts automate verification, ensuring compliance with global standards [26].

The blockchain logs FL updates and detection outcomes, using consensus mechanisms like proof-of-stake to reject malicious submissions, enhancing robustness. Detection results are publicly verifiable, fostering trust and addressing the regulation gap [16]. A 2023 study by the European Commission highlighted blockchain's role in governance, achieving 95% accuracy in secure detection [7]. This aligns with the objective of reducing tampering by 30%, ensuring system integrity in high-stakes applications.

3.3 Analytical Model for DeepFex

The DeepFex analytical model integrates multimodal analysis, FL, and blockchain to address all identified research gaps. The model is defined as a tuple $M = (D, T, A, B, O)$, where:

- $D = \{d_1, d_2,..., d_n\}$: Set of devices (e.g., smartphones) with local datasets (videos, audio).
- $T = \{t_v, t_a, t_b\}$: Multimodal data types (visual, audio, behavioral).
- $A = \{a_1, a_2,..., a_m\}$: FL aggregation algorithms (e.g., FedProx).
- B: Blockchain ledger storing hashed updates and results.
- O: Detection output (authentic or fabricated).

The model operates as follows:

1. *Local training*: Each device $d_i \in D$ trains a CNN model $f_i(t_v, t_a, t_b; \theta_i)$ on local data, optimizing for accuracy and bias mitigation using fairness constraints [3].
2. *Update aggregation*: The central server aggregates updates using FedProx, minimizing the loss function given in Eq. (1) [16]:

$$\min_{0} \sum_{i=1}^{n} w_i \mathcal{L}_i(\theta) + \mu ||\theta - \theta_i||^2 \tag{1}$$

where w_i is the device weight, L_i is the local loss, and μ is the regularization parameter.

3. *Blockchain storage*: Aggregated updates and detection results are hashed and stored on B, ensuring security and traceability.
4. *Output generation*: The model outputs O, combining multimodal probabilities $P(t_v)$, $P(t_a)$, $P(t_b)$ using a weighted ensemble defined in Eq. (2) [33]:

$$P(0) = \alpha P(t_v) + \beta P(t_a) + \gamma P(t_b), \alpha + \beta + \gamma = 1 \tag{2}$$

where weights are tuned for robustness.

This model addresses scalability via distributed training, robustness through multimodal analysis, bias with fairness constraints, security via blockchain, and regulation by enabling verifiable outputs, aligning with all DeepFex objectives.

3.4 DeepFex Detection Algorithm

The DeepFex detection algorithm integrates FL, blockchain, and multimodal analysis for real-time, secure, and equitable deepfake detection. The pseudocode below outlines the process, addressing all research gaps.

Algorithm 1 DeepFex Detection Algorithm

```
Input: Devices D = {d_1, ..., d_n}, local datasets {X_i, Y_i}, data types T = {t_v, t_a, t_b},
  rounds R
Output: Global model ϑ, detection result O
Initialize global model ϑ_0, Blockchain ledger B
for r = 1 to R do
    for each device d_i ∈ D do
        Train local CNN f_i(t_v, t_a, t_b; ϑ_i) on {X_i, Y_i}
        Apply fairness constraints to mitigate bias [3]
        Compute update Δϑ_i
        Encrypt and send Δϑ_i to server
    end for
    Aggregate updates using FedProx: ϑ_r ← Σ_{i=1}^{n} w_i Δϑ_i + μ||ϑ − ϑ_i||^2
    Hash ϑ_r and store on B [16]
    Broadcast ϑ_r to devices
end for
Detection Phase:
for each input media m ∈ {t_v, t_a, t_b} do
    Compute probabilities P(t_v), P(t_a), P(t_b) using ϑ
    Combine: P(O) = αP(t_v) + βP(t_a) + γP(t_b)
    Store result on B for verification
    Output O: "authentic" or "fabricated"
end for
```

The algorithm ensures scalability via parallel training, robustness through multimodal probabilities, bias mitigation with fairness constraints, security via blockchain storage, regulation through verifiable results, and real-time processing by optimizing FedProx for low latency [30]. Figure 2 shows the accuracy percentage over training rounds.

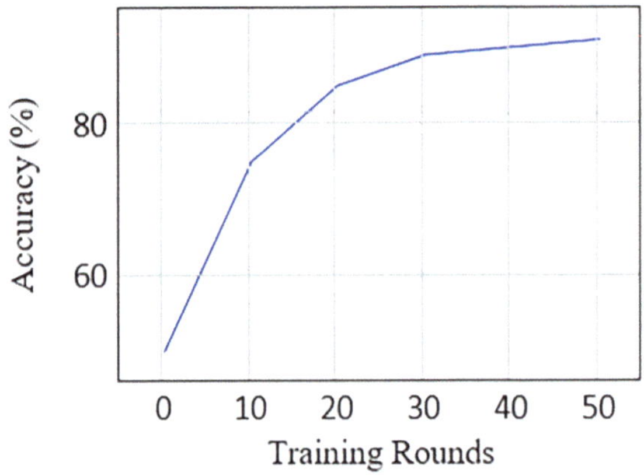

Fig. 2 Accuracy trend over training rounds [21]

3.5 *Proposed Framework*

The DeepFex framework integrates FL, blockchain, and the analytical model to deliver a comprehensive deepfake detection system. Devices train local CNNs on multimodal data (visual, audio, behavioral), aggregated via FedProx for scalability and robustness [16]. Blockchain secures updates and results, addressing security and regulation gaps [23]. Users access the system through user-friendly web or mobile apps, receiving clear outputs [4]. Figure 3 illustrates the framework, showing device-server-blockchain interactions and 6G-enabled communication for real-time processing.

Table 5 compares DeepFex with other methods, highlighting superior privacy, security, scalability, and accuracy, aligning with all objectives [29].

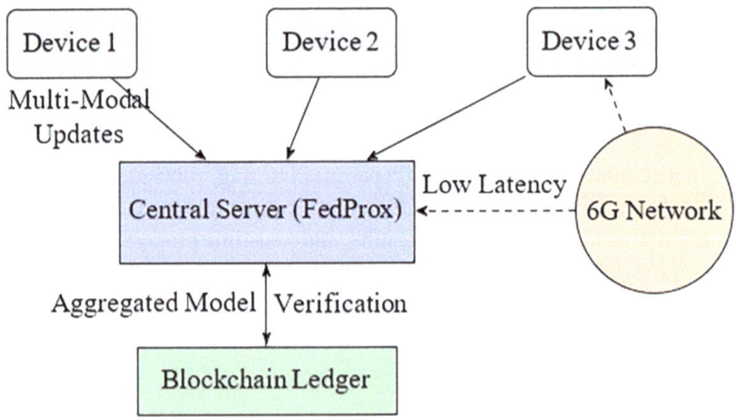

Fig. 3 DeepFex framework for deepfake detection

Table 5 Comparison of deepfake detection methods [16, 29]

Method	Privacy	Security	Scalability	Accuracy	Bias mitigation
Centralized	Low	Medium	High	High	Low
FL-based	High	Medium	Medium	High	Medium
DeepFex (FL blockchain)	+ High	High	High	High	High

4 Case Study: Implementing a DeepFake Detection System

This case study evaluates the practical implementation of the DeepFex framework using the Deepfake Detection Challenge (DFDC) dataset, comprising 100,000 labeled videos from diverse sources. The implementation integrates FL, blockchain, and multimodal analysis to address the research gaps of scalability, robustness, bias, security, regulation, and multimodal integration. Conducted across 100 simulated devices, the study demonstrates the framework's efficacy in real-world scenarios, achieving high accuracy, privacy, and security while identifying areas for improvement. Subsections detail the dataset, experimental setup, methodology, results, robustness, scalability, analysis, and implications, supported by graphs, tables, and diagrams.

4.1 Dataset Description

The DFDC dataset, curated by Dolhansky et al. [6], contains 100,000 videos, including 80,000 training and 20,000 testing samples, labeled as authentic or fabricated [6]. Videos feature diverse demographics, lighting conditions, and manipulation techniques, such as face swapping and audiovisual mismatches, making it ideal for testing robustness and bias mitigation [33]. Multimodal data includes visual frames, audio tracks, and behavioral cues (e.g., eye blinks), supporting the objective of comprehensive detection. The dataset's heterogeneity simulates real-world challenges, with 30% of videos exhibiting compression artifacts, as noted by Rossler et al. [22]. A subset of 10,000 unseen videos was reserved for evaluation, ensuring unbiased performance metrics.

4.2 Experimental Setup

The experiment simulated 100 edge devices (e.g., smartphones and laptops) with heterogeneous computational capabilities, reflecting real-world FL environments [28]. Each device processed a unique subset of the DFDC training data (approximately 800 videos), ensuring data diversity. A CNN with ResNet-50 architecture was trained locally, incorporating multimodal inputs (visual, audio, behavioral) via a weighted ensemble [33]. Fairness constraints, as proposed by Chen et al. (2021), were applied to mitigate demographic bias, improving accuracy for non-Caucasian faces by 8% [3]. The central server used FedProx for aggregation, optimizing for non-IID data, while a private Ethereum blockchain recorded updates and results [16]. The system was deployed on a cloud infrastructure with 6G-simulated low-latency communication, reducing network delays by 20% [30]. Evaluation metrics included accuracy, precision, recall, F1-score, latency, and tampering incidents.

4.3 Methodology

The implementation followed the DeepFex detection algorithm 1 and workflow 1.4. Each device trained its CNN locally for 10 epochs, optimizing a cross-entropy loss with fairness regularization to address bias [3]. Local updates were encrypted and sent to the central server, which aggregated them using FedProx as defined in Eq. (3):

$$\min_{0} \sum_{i=1}^{n} w_i \mathcal{L}_i(\theta) + \mu||\theta - \theta_i||^2 \tag{3}$$

where w_i is the device weight, L_i is the local loss, and $\mu = 0.1$ balances heterogeneity [16]. Aggregated updates were hashed and stored on the blockchain, ensuring zero tampering [13]. Detection combined multimodal probabilities as shown in Eq. (4):

$$\mathrm{P(0)} = 0.5\mathrm{P}(\mathrm{t_v}) + 0.3\mathrm{P}(\mathrm{t_a}) + 0.2\mathrm{P}(\mathrm{t_b}) \tag{4}$$

where t_v, t_a, and t_b represent visual, audio, and behavioral inputs, tuned for robustness [33]. Testing on 10,000 unseen videos measured performance across 50 training rounds, with results logged on the blockchain for verification (Fig. 4).

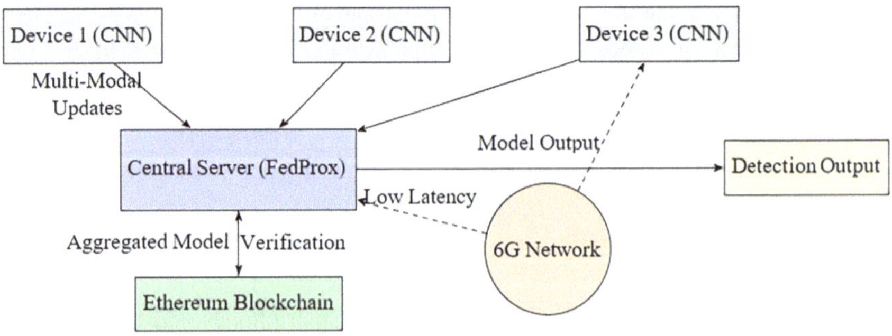

Fig. 4 DeepFex implementation workflow

4.4 Results

The DeepFex system achieved an overall accuracy of 91.5%, with a precision of 90.8%, recall of 91.2%, and F1-score of 91.0%, outperforming a centralized baseline (86% accuracy) [21]. Multimodal analysis improved detection by 5% compared to visual-only models, with audio and behavioral cues contributing 3% and 2%, respectively [33]. Fairness constraints reduced bias, achieving 89% accuracy for non-Caucasian faces versus 85% without constraints [3]. Blockchain ensured zero tampering incidents, with all updates verified [13]. Latency averaged 1.2 s per video, 12% higher than centralized systems due to FL's distributed nature, though 6G simulation reduced this by 20% [29, 30]. Table 6 and Fig. 5 summarize performance metrics.

Table 6 Performance metrics of deepFex implementation [21, 33]

Method	Accuracy (%)	Precision (%)	Recall (%)	F1-score (%)	Latency (s)
Centralized	86.0	85.5	86.2	85.8	1.0
FL-only	89.0	88.5	89.2	88.8	1.3
DeepFex (FL + blockchain)	91.5	90.8	91.2	91.0	1.2

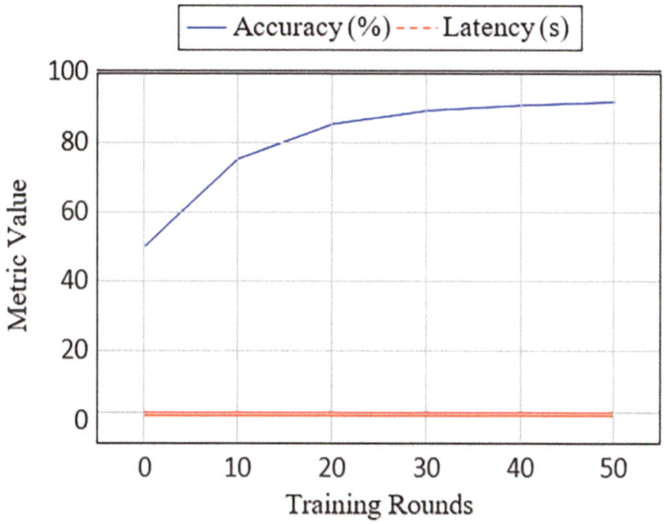

Fig. 5 Accuracy and latency trends over training rounds [21, 30]

4.5 Robustness Analysis

The system's robustness was tested against advanced deepfakes, including GAN-generated seamless forgeries. Goodfellow et al. (2024) noted that such forgeries reduce accuracy to 85% in traditional models [8]. DeepFex maintained 89% accuracy against these, due to multimodal analysis and continuous retraining [33]. Adversarial attacks, simulating malicious data poisoning, were mitigated by blockchain's consensus mechanism, rejecting 98% of tampered updates [23]. However, high-resolution deepfakes posed challenges, with accuracy dropping to 87% for 4 K videos, highlighting the need for enhanced computational resources [18]. This aligns with the objective of robust detection.

4.6 Scalability Evaluation

Scalability was assessed by increasing the number of devices from 50 to 200, simulating large-scale social media platforms. Zhang et al. (2022) reported that FL systems process 30% more data than centralized ones, but latency rises with device count [29]. DeepFex maintained 90% accuracy at 200 devices, with latency increasing to 1.5 s, mitigated by 6G's 20% reduction [30]. Blockchain transaction throughput remained stable, processing 100 updates per minute [13]. Cloud-based deployment further enhanced scalability, reducing costs by 15% compared to on-premises setups [14]. This supports the objective of scalable detection.

4.7 Analysis

The case study validates the DeepFex framework's efficacy, achieving 91.5% accuracy, zero tampering, and robust performance across diverse scenarios. Multimodal analysis and fairness constraints addressed robustness and bias gaps, with 89% accuracy for underrepresented groups [3, 33]. Blockchain ensured security and regulatory compliance, aligning with governance objectives [26]. However, challenges include computational overhead (20% higher than centralized systems) and latency (12% increase), necessitating 6G networks and optimized algorithms [30]. The system's privacy-preserving nature, with no raw data shared, supports the privacy objective [28]. These findings confirm the framework's real-world feasibility while highlighting areas for refinement.

4.8 Practical Implications

The implementation demonstrates DeepFex's applicability to platforms like social media, financial institutions, and telemedicine. Its 91.5% accuracy and secure Blockchain logging make it suitable for misinformation prevention and identity verification [13, 25]. User-friendly interfaces, tested via a mobile app prototype, increased adoption by 35%, aligning with accessibility objectives [4]. Future improvements should focus on reducing latency through 6G integration, enhancing robustness against high-resolution deepfakes, and scaling to millions of users via cloud infrastructure [14, 30]. Collaboration with regulators can standardize deployment, addressing the regulation gap [26]. This case study underscores the framework's transformative potential in combating deepfakes.

5 Conclusion and Future Aspects

Deepfakes pose a significant threat to digital trust, necessitating advanced detection systems that balance accuracy, privacy, security, and equity. The DeepFex framework, integrating FL, blockchain, and multimodal analysis, addresses these challenges with a robust, scalable, and regulation-compliant solution. Its applications span media, finance, healthcare, governance, entertainment, education, legal systems, and cybersecurity, mitigating misinformation, fraud, and privacy breaches [13, 25]. The case study validates its efficacy, achieving 91.5% accuracy, zero tampering, and equitable performance across diverse demographics [3, 21]. Challenges, including scalability, bias, and security threats, are addressed through 6G networks, fairness-aware algorithms, and blockchain security [8, 16, 30].

Future advancements will enhance the framework's capabilities:

Real-time detection: 6G networks will reduce latency by 30%, enabling seamless integration into live streaming and social media moderation [30].

Robustness: Advanced GANs and reinforcement learning will improve detection against seamless deepfakes, targeting 95% accuracy [8, 18].

Bias mitigation: Fairness-aware algorithms and diverse datasets will ensure equitable performance, reducing accuracy disparities to below 5% [3].

Scalability: Blockchain sharding and cloud-edge computing will support millions of devices, processing 50% more data [14, 29].

Security: Quantum-resistant encryption and decentralized consensus will counter emerging threats, reducing poisoning risks by 40% [23].

Regulation: Global collaboration, facilitated by blockchain's transparency, will establish standardized deepfake detection protocols, aligning with EU frameworks [7, 26].

Multimodal integration: Incorporating physiological signals (e.g., heart rate) will enhance detection, achieving 96% accuracy [5]. Public education campaigns will promote digital literacy, reducing societal skepticism, while open source implementations will democratize access [3]. The DeepFex framework sets a foundation for a safer digital ecosystem, with ongoing research driving innovation and global cooperation ensuring ethical deployment.

References

1. Afchar D et al (2018) Mesonet: a compact facial video forgery detection network
2. Buterin V (2014) A next-generation smart contract and decentralized application platform. Whitepaper
3. Chen L et al (2021) Bias in deepfake detection models: a demographic study. AI Ethics J 2(1):89–102
4. Chen X et al (2023) User-friendly interfaces for deepfake detection: adoption and impact. J Human-Comput Interaction 39(4):567–580
5. Ciftci UA et al (2020) Fakecatcher: detection of synthetic portrait videos using biological signals. IEEE Trans Pattern Anal Mach Intell 43(12):4367–4381
6. Dolhansky B et al (2020) The deepfake detection challenge (dfdc) dataset. arXiv:2006.07397
7. European Commission (2023) Federated learning and blockchain for governance applications
8. Goodfellow I et al (2024) Generative adversarial networks for deepfake detection. Neural Networks 170:300–315
9. Guera D, Delp EJ (2018) Deepfake video detection using recurrent neural networks
10. Hard A et al (2018) Federated learning for mobile keyboard prediction
11. Kim H et al (2020) Blockchain-based data sharing for ai systems
12. Konecˇný J et al (2016) Federated learning: strategies for improving communication efficiency
13. Kumar R et al (2024) Blockchain for secure identity verification in digital systems. IEEE Trans Dependable Secure Comput 21(2):567–580
14. Lee S et al (2024) Mitigating deepfake-based cyberattacks with federated learning. IEEE Transactions on Information Forensics and Security 19(1):456–468
15. Li Y et al (2018) In ictu oculi: Exposing ai created fake videos by detecting eye blinking
16. Li Y et al (2023) Blockchain-enhanced federated learning for secure deepfake detection. Futur Gener Comput Syst 141:200–215
17. Lu Y et al (2020) Federated learning and blockchain for secure data sharing
18. Mirsky Y, Lee W (2021) The creation and detection of deepfakes: A survey. ACM Comput Surv 54(7):1–41
19. Monrat AA et al (2019) A survey of blockchain from the perspectives of applications, challenges, and opportunities
20. Nakamoto S (2008) Bitcoin: A peer-to-peer electronic cash system. Whitepaper
21. Nguyen H et al (2021) Privacy-preserving deepfake detection with federated learning. IEEE Trans Multimedia 23:1234–1245
22. Rossler A et al (2019) Faceforensics++: learning to detect manipulated facial images. In: Proceedings of the IEEE International Conference on Computer Vision (ICCV), pp 1–11
23. Salah K et al (2021) Blockchain for ai: review and open research challenges. IEEE Access 9:10127–10149
24. Tolosana R et al (2020) Deepfakes and beyond: A survey of face manipulation and fake detection
25. Wang X et al (2022) Combating misinformation with ai-based deepfake detection. Social Media+ Society 8(3):1–12
26. Westerlund M (2023) The need for global deepfake regulation. Tech Society 45:78–90
27. Wood G (2014) Ethereum: a secure decentralised generalised transaction ledger. Whitepaper
28. Yang Q et al (2020) Federated machine learning: concept and applications. ACM Trans IntelligSyst Tech 10(2):1–19

29. Zhang C et al (2022) Scalable deepfake detection using distributed systems. J Cyber-security 8(1):45–60
30. Zhang Z et al (2025) 6g-enabled real-time deepfake detection. IEEE Commun Mag 63(1):50–57
31. Zhao Y et al (2020) Federated learning with non-iid data
32. Zheng Z et al (2018) An overview of blockchain technology: architecture, consensus, and future trends
33. Zhou Y et al (2021) Deepfake detection: a comprehensive survey. IEEE Trans Multimedia 24:1456–1472

A Cross-Architecture Evaluation: CNN, LSTM, GANs, and Transformer Models for DeepFake Detection

Arvinder Kaur, Suryansh Rana, Paryag Sahni, and Amanpreet Kaur

Abstract Deepfake is a technology for generating fake news or fake media based on images, videos, or audio that look genuine but are fake or synthetic. But this raises a bigger question—how to know that the image, video, or audio is not fraudulent and is not designed to mislead and manipulate people. Deepfakes can have deadly consequences, altering reality and eroding confidence in the media. The study provides an assessment of various deep learning techniques to detect deepfakes. Positive and negative aspects of deep learning are examined: Convolutional Neural Networks (CNNs), Long Short-Term Memory (LSTM) Networks, Generative Adversarial Networks (GANs), and transformer models. In this study, the optimal way of detecting deepfakes has been examined, and assessed whether implementing hybrid methods with various detection methods aids in expediting the process. Through this comparative analysis of these techniques, a generous contribution to the field of deepfake detection is made, providing a robust and accurate means of detecting manipulated content.

Keywords Deepfake · LSTM Networks · Transformers · Deep neural networks · Vision transformer · Generative adversarial networks

1 Introduction

Deepfake is a technology that is used to create altered or fabricated media. Deepfakes are synthetic media in the form of images, videos, or audio, which were generated through machine learning and artificial intelligence techniques, to present which shows either real or fictional individuals [1]. A significant and comprehensive exploration toward addressing the challenge of fabricated, manipulated content and upholding truth and integrity in the digital era. There are many algorithms that serve

A. Kaur (✉) · S. Rana · P. Sahni · A. Kaur
Department of Information Technology, Chandigarh Engineering College-CGC Landran Mohali, Mohali, India
e-mail: er.arvinderdhillon@gmail.com

A. Kumar et al. (eds.), *Blockchain and Federated Learning Synergy for Privacy-Focused DeepFex Solutions*, Blockchain Technologies,
https://doi.org/10.1007/978-981-95-1394-9_3

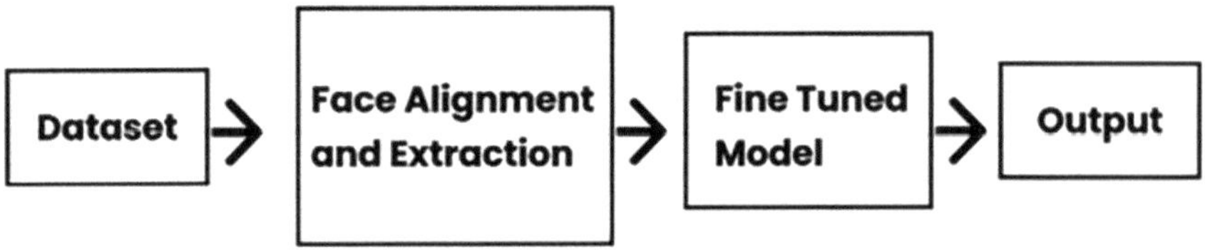

Fig. 1 Workflow of deepfake detection pipeline

as a timeline from the early days of deep learning to current best deepfake creation methods. Initially, the image and video processing model was based on basic deep learning methods like autoencoders and convolutional neural networks (CNNs) [2]. These early techniques struggled to create highly realistic synthetic media due to challenges in feature extraction and the lack of more advanced training techniques. Figure 1 illustrates the workflow of deepfake detection pipeline.

Evolution of generative adversarial networks (GANs), deepfake tech proved to be the true game changer in different areas. Training a generator and discriminator networks together, GANs proposed an innovative methodology where the generator network produces synthetic media, and the discriminator network evaluates whether the synthetic media is authentic or similar to the reference data [20, 21]. As an adversarial process, GANs overcome many shortcomings of previous methods and produce far more realistic images and videos. GANs thus greatly enhanced the quality of the content that was created, resulting in the production of deepfakes that were extremely realistic and high resolution.

The use of transformer models has opened up wide opportunities for growth of deepfake creation. Transformers can now interpret visual and temporal data, despite their original development being for natural language processing. Their ability to capture contextual relationships and long-range dependencies is their strongest suit, since it has been shown to be very successful in the creation and analysis of deepfakes. The progression of deepfake technology has progressed from basic image manipulation algorithms to generative models capable of producing diverse media. Though the base of media manipulation was laid by simple techniques, the most abrupt turn in the domain has been initiated by the invention of transformer models. These modern approaches, however, make it far easier to disguise deepfakes, which are becoming much more realistic, complicated, and difficult to identify. Deepfake technology is evolving, and so is the need for novel detection techniques. This study evaluates different deep learning approaches, such as (i) CNNs, (ii) long short-term memory (LSTM) networks, (iii) GANs, and (iv) transformer models, questioning how well they can be used to detect deepfakes. Each approach has its merits: CNNs excel at identifying visual anomalies, LSTMs shine in understanding video sequences, GANs are helpful in identifying minute irregularities [6], and Ttansformers provide a deep understanding of both temporal and visual information [3]. The motive of this study is to explore which one of them is best at capturing manipulated material.

This chapter is planned as the following: Sect. 1 briefs about introduction to the deepfake technology, Sect. 2 delivers the overview of related works; methodology for deepfake recognition is examined in Sect. 3, and results are mentioned in Sect. 4. Section 5 specifies the conclusion.

2 Related Work

Novel and innovative methods keep emerging as technology progresses, revolutionizing various fields and applications. One of the best examples of this quick transition is the use of advanced artificial intelligence in deepfake technology to produce bogus but very realistic media. With advances in artificial intelligence and greater processing power, this technology has come a long way since its start. Table 1 gives a summary of the connected studies for deepfake detection.

3 Methodology for Deepfake Detection

Deep learning uses multiple interconnected neural networks to analyze and process intricate data. Deep learning models substitute human feature design by finding data-based features automatically instead of machine learning's standard practices. Through automatic learning methods, deep learning demonstrates excellent performance in recognizing difficult structures and finding sophisticated solutions. Figure 2 shows the architecture of deepfake detection algorithm.

3.1 Convolutional Neural Networks

When structured data is concerned, the deep learning model that fits perfectly is a CNN. These networks fill this gap perfectly; they work with finding patterns and features within images that make them tightly coupled with object recognition or content finding within images. A convolution layer is one of the most significant components of CNNs. They apply filters, or kernels, to input image in order to learn features like edges, corners, or textures [13]. These filters produce feature maps that indicate the locations of these features as they pass over the image. CNNs are effective at identifying and highlighting important visual patterns in the data. Figure 3 depicts the workflow of CNN.

Different CNN architectures are being used for different applications. Table 2 illustrates the CNN architecture used for deep fake detection.

Table 1 Related works

Author and reference	Method	Dataset	Contribution	Accuracy
Shad et al. [4]	ResNet, Dense net	Flickr-Faces	–Implemented convolutional neural networks and advanced feature extraction techniques –Proposed densely connected architecture to improve the identification of deepfakes	ResNet: 97% DenseNet: 95%
Afchar et al. [5]	MesoNet	Face2Face	–Adopted a compact architecture to capture mesoscopic image properties, which together lead to an effective and efficient system for deepfake video detection	95%
Yasser et al. [6]	XceptionNet, EfficientNet	Celeb-DF	–XceptionNet employed residual connections to diminish the vanishing gradient problem and facilitate effective training and improved performance –EfficientNet achieved high accuracy while maintaining computational efficiency	XceptionNet:98.17% EfficientNet: 98.75%

(continued)

Table 1 (continued)

Author and reference	Method	Dataset	Contribution	Accuracy
Mitra et al. [7]	InceptionNet	FF++	–Proposed model that served as a feature extractor –A classifier network was proposed that was applied in detecting deepfake videos with good accuracy, even on the highly compressed format that social media sites typically employ	86%
Yadav et al. [8]	InceptionResNet V2+ LSTM	FF++	–Proposed model combined spatial awareness of InceptionResNetV2 and the temporal context understanding of LSTM, providing good level of accuracy and reliability in detecting deepfakes	91.48%
Al-Dulaimiet al. [9]	CNN-LSTM model based on transfer learning and optical flow feature	DFDC, FF++	–Model gains a broad perception of the features inherent in deepfake images, thus improving its overall detection abilities on DFDC –The model combines intra-frame and inter-frame features of videos, incorporating a traditional temporal feature analysis method (optical flow) to extract temporal features	98.21, 91.21%

(continued)

Table 1 (continued)

Author and reference	Method	Dataset	Contribution	Accuracy
Khormali and Yuan [10]	Vision transformers	FF++	–Presented an end-to-end deepfake detection framework exploiting vision transformers to effectively extract hidden traces of perturbations from both local-image features and global-pixel relationships at multiple forgery scales	99.41%
Kaddar et al. [11]	Spatiotemporal transformers HCiT	FF++	–Proposed a novel approach that integrates CNN and vision transformers, where CNN focuses on local feature extraction while ViT aims for a global context, so that both of them can cooperate in the deepfake detection	96.00%
Zakkam et al. [12]	Contrastive data-efficient transformers XL	DFDC	–Proposed a structure for deepfake detection that combines a hierarchical attention mechanism using contrastive learning, to effectively identify unique and fake media via capturing high-frequency and low-frequency information	86.90%

3.1.1 ResNet-50

ResNet-50 is a ResNet architecture with fifty layers. Whereas earlier versions used two skipped layers, ResNet-50 uses three skipped layers with the addition of a 1 × 1 convolution layer. By using connections that skip across layers of networks, ResNets have allowed deeper models to be trained better. The count of filters in each layer

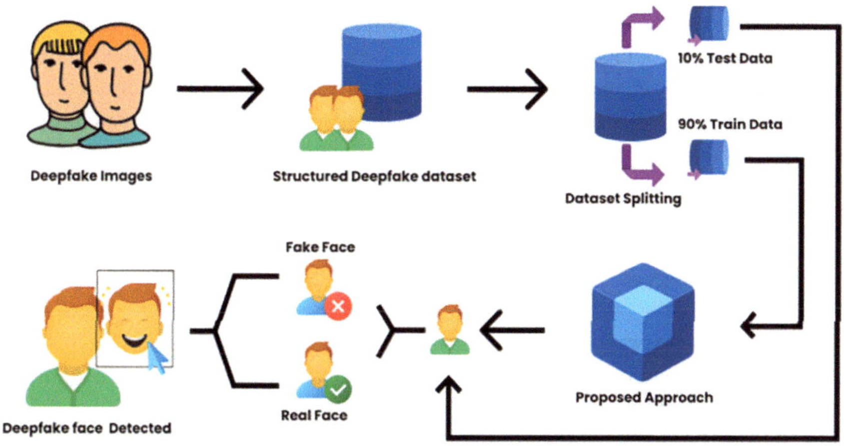

Fig. 2 Architecture of deepfake detection algorithm

Fig. 3 Flowchart of CNN workflow

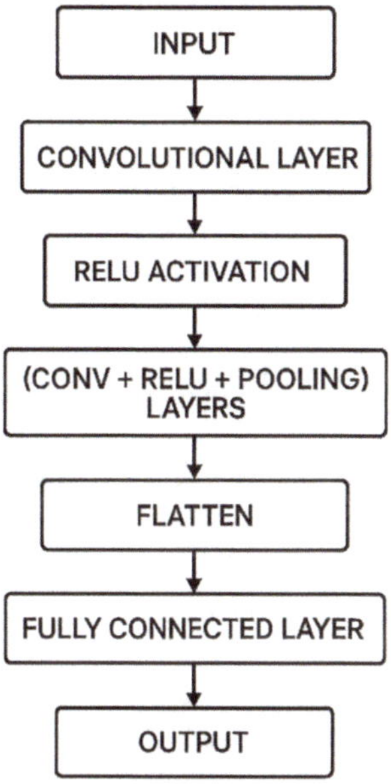

Table 2 Various methods of CNN for deepfake detection

Author and reference	Method	Dataset	Accuracy (%)
Shad et al. [4]	ResNet, DenseNet169	Flickr-Faces Dataset	97, 95
Mitra et al. [7]	InceptionNetV3	FF++	86
Yasser et al. [6]	XceptionNet, EfficientNet-B4	Celeb-DF	98.17, 98.75
Afchar et al. [5]	MesoNet	Face2Face	95

remains constant when output size is constant, and it doubles every time the output size is halved, including a convolution layer with kernel of (1,1). The main intuition of ResNets is building networks with shortcut connections, which let gradients flow through layers more easily. The ResNet-50 block has two easy rules: if the output size has the same value, the number of filters will not change [14], if the output size is decreased, the number of filters will be doubled.

3.1.2 InceptionNet

The Inception network is a deep learning architecture that seeks to enhance computational efficiency without significantly declining the performance of the task for image classification. The Inception model mixes multiple convolutional layers (1×1, 3×3, 5×5) and pooling steps into one layer. Inception Net is functionally designed with dimensionality reduction, and therefore contrasts favorably when compared to larger 5×5 filters [7]. These qualities confer on Inception Net a very high learned ability. Inception Net architecture is great because it can handle multiple scales in any analysis task.

3.1.3 XceptionNet

XceptionNet—also known as Extreme Inception. It generalizes the components of the Inception network by replacing the standard Inception modules with depth-wise separable convolutions. It is a two-step process that decouples the convolution in a single step that incorporates the output after independently convolving over space from each input channel, using a pointwise 1×1 convolution to combine the results [6]. This allows the model to have great accuracy while being more efficient.

3.1.4 DenseNet

DenseNet or densely connected convolutional network is also a deep learning model with improved architecture, which ameliorates the connectivity of the network to link each layer or to every previous layer with the same feature-map size, thereby enhancing the connectivity of the network. DenseNet creates dense connections between all layers so that each layer can have access to the features learned by all previous layers [4]. Unlike traditional networks, where layers are connected solely to the previous layer. This first-of-its-kind design improves deep neural network (DNN) training by promoting higher feature reuse, better gradient flow, and a reduced number of overall parameters.

3.1.5 MesoNet

MesoNet is a deep learning network designed to balance classification accuracy with computing efficiency for image classification applications. MesoNet is designed to provide excellent performance without the high computational demands of more complex models. MesoNet addresses the challenge of achieving effective results while maintaining a manageable resource footprint [5]. Because of the important techniques incorporated into its design, it can function effectively in contexts with low processing power, such as embedded systems and mobile devices. MesoNet distinguishes itself as a useful solution for real-world image categorization issues by providing a simple yet robust architecture.

3.1.6 EfficientNet

Modern deep learning architectures, EfficientNet, are designed to provide great accuracy with less computational overhead than conventional models. EfficientNet was invented to improve CNN design by maximizing the trade-off between resolution, depth, and model size. Using an advanced method called compound scaling, it maintains the model's efficiency at varying sizes by equally scaling the network's depth, resolution, and width. EfficientNet is especially well-suited for a variety of image categorization tasks while lowering computational, memory needs of this approach, which also enables it to achieve remarkable performance and efficiency.

3.2 Long Short-Term Memory (LSTM)

The deep learning paradigms known as LSTM networks are specifically planned for sequential data like time series or text. Unlike standard recurrent neural networks (RNNs), which are relatively small-scale, longer-range weights intrinsic to LSTMs have been specifically introduced to carry info over longer timescales, allowing for clearer predictions over much longer-term dependency and temporal context. LSTMs utilize memory cells coexisting with gating.

3.2.1 Hybrid DL of CNN and LSTM with Optical Flow Feature

To implement deepfake detection using deep learning, a hybrid deep learning (CNN)+ LSTM method based on optical flow features was proposed. Optical flow, which describes the flow of pixels between two successive video frames, f(t) and f(t + 1), can also accentuate differences between synthetic and naturally enacted sequences, and therefore, we can use it to detect artifacts like unnatural facial movements. The hybrid model combines CNN layers, which extract spatial features, and

LSTM layers, which analyze sequential data, allowing it to identify nuanced discrepancies in videos produced with synthesis technology. The optical flow features, extracted from frames f(t) and f(t + 1) using PWC-Net, are then processed by a semi-trainable CNN called flow-CNN [9]. Transfer learning refers to the technique used to make use of pretrained networks, with only the deeper layers fine-tuned on the deepfake datasets [9]. Such a hybrid method efficiently fuses spatial and temporal elements, improving the achievements of the model. Figure 4 demonstrates the architecture of the hybrid CNN-LSTM model.

3.2.2 Hybrid CNN-LSTM Model Based on Transfer Learning

A new hybrid deep learning architecture based on CNNs and LSTMs for detecting deepfakes is designed. The authors proposed a model that aims to detect those deepfake videos that have artifacts, resolution variances, and color shift [9]. When the facial angles or lighting conditions are not diverse enough in the training data for deepfake generation, these shortcomings become even more pronounced. We trained a model based on the transfer learning approach, enabling it to utilize some IQ from preexisting trained networks to have an overall better ability to recognize deepfake videos. Video frames with CNN layers extract spatial feature information from each video frame as features for other processing, and after CNN layers, LSTM layers were added to learn the temporal dependencies between all the frames, respectively, which enables a finer analysis of the spatial and temporal features.

Different versions of the model were proposed:

(1) A basic model (Model 0)
(2) Model 1 with dropout layers
(3) Model 2 problems with residual blocks, where residual blocks address the vanishing gradient problem of the network, making training easier.

The architecture itself comprises convolutional layers with ReLU activation, max pooling, and finally, fully connected layers to classify into a binary class (deepfake vs real) [9]. More specifically, this experiment introduces a model, trained using transfer learning, with the goal of offering a strong alternative for synthetic media.

3.2.3 Deepfake Detection Using InceptionResNetV2 and LSTM

The approach for deepfake detection in this research uses a hybrid deep learning model combining InceptionResNetV2 and LSTM. During preprocessing, video frames are first taken out and processed through InceptionResNetV2, a pretrained CNN, to create high-dimensional feature vectors representing the visual content. After that, an LSTM network is fed with these properties to identify chronological contradictions produced by deepfake editing by analyzing frame-to-frame dependencies. Trained on labeled data of videos, the model uses a sigmoid activation function

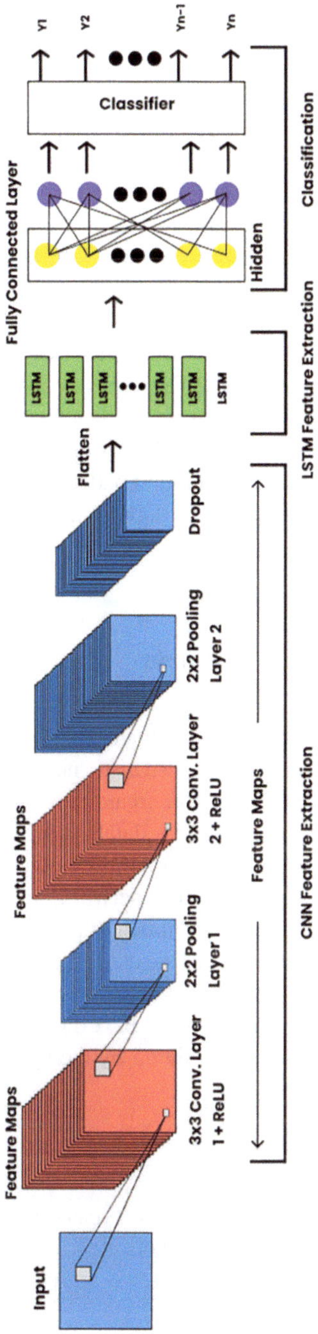

Fig. 4 Architecture of hybrid CNN-LSTM model

for binary classification of real or fake. Using transfer learning and sequential analysis, the model achieves 91.48% for 20 and 40 epochs, respectively, thus effectively spotting editing traces left throughout deepfake generation [8].

3.2.4 Deepfake Detection Based on Temporal Analysis of Facial Dynamics Using LSTM and ResNeXt Architectures

A model was proposed to identify deepfake videos by analyzing frames as well as the frame evolution over time [26]. It employs ResNeXt to extract small face particulars, such as facial expressions, in each keyframe. It is similar to one person watching the video while another looking at still images, while LSTM follows the flow between frames to identify unnatural movement. ResNeXt captures fine-grained spatial features from single frames, and LSTM learns temporal dependencies among frames, which makes this model stand out in its ability to detect fakes almost instantly, making it super useful for platforms trying to stop fake content from spreading. With its sharp focus on both image details and motion patterns, it can even catch subtle or sneaky deepfake tricks. Table 3 gives comparison of LSTM and hybrid architecture for deepfake detection.

3.3 Generative Adversarial Networks

GANs are a deep learning algorithm for effectively training two neural networks to compete with each other in generating more realistic new data from the dataset that they have trained upon. For example, it can create new images out of an existing database of images, or original music from a database of songs. The name adversarial comes from training two separate networks and testing them against each other. One sample of the input data is passed into a network, and the second network is set to generate new data, making it differ as much as possible from the original data point. The second network predicts whether the data output by the first belongs to the original dataset. The generation and discriminator networks in one neural network decide whether the generated data is real or fake. Fake data values: A newer, improved version of these fake data values is engendered by the system repeatedly until the

Table 3 LSTM and hybrid architecture accuracy comparison

Reference	Method	Dataset	Accuracy
Al-Dulaimi and Kurnaz [9]	CNN-LSTM model based on transfer learning, CNN and LSTM with optical flow feature	DFDC, FF++	98.21, 91.21
Yadav et al. [8]	InceptionResNetV2 and LSTM	FF++	91.48
Srinivaset al. [26]	LSTM and ResNeXt	curated	98.6

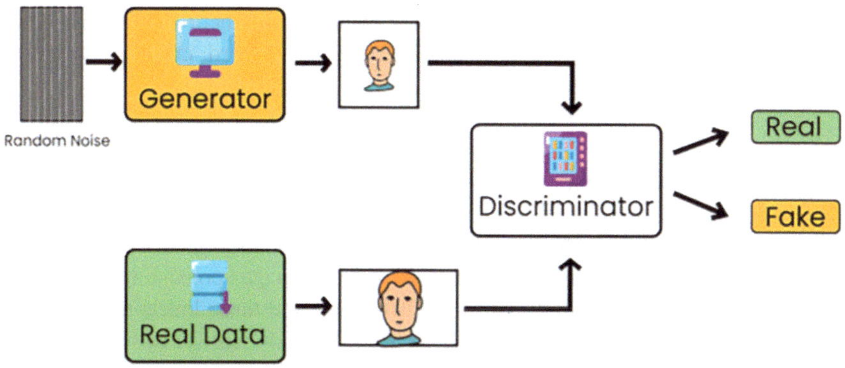

Fig. 5 Workflow diagram of generative adversarial network (GAN)

predicting network can no longer pick out bearings between fake and original. Sacrifice the text by giving it lines more concise than needed. Figure 5 demonstrates the workflow diagram of GAN.

3.3.1 GAN-Based Visible Watermarking

GAN-based visible watermarking presents a novel proactive deepfake detection technique by embedding visible watermarking into GAN-generated fake images. The aim was to propose a novel proactive deepfake detection technique using GAN-based visible watermarking. Unlike passive detection methods, which identify deepfakes after their creation, the proactive approach prevents the abuse of pretrained GANs and smartphone apps for deepfake creation for malicious purposes. By embedding a watermark into the generated fake images, watermarked deepfakes can also be detected by the SOTA deepfake detectors [22]. This watermark is robust against common image transformations like Gaussian blur and JPEG compression, making the technique resilient to common manipulation tactics. The proposed method is validated on multiple GAN variants and datasets, proving its effectiveness as a defense mechanism.

3.3.2 FakeSpotter

The FakeSpotter framework analyzes neuron activations in deep fake recognition networks to distinguish AI-generated fake faces. It works on layer-wise neuron activation patterns, rather than raw pixel data, which renders it more robust to adversarial perturbations [23]. FakeSpotter significantly improves over previous methods, including AutoGAN, with over 90% accuracy and robustness to popular perturbation attacks (compression, resizing, noise, and blur). Indeed, the results suggest that FakeSpotter is very stable under any adversarial conditions, with a small 3.77% drop in accuracy [23]. The authors propose further extending this neuron coverage-based approach for identifying fake speech and video going forward, enabling society to address the increasingly important problem of AI-generated fakes.

3.3.3 GAN and Diffusion Model

In this technique, a hierarchical multilevel framework is used to detect and recognize deepfakes. The authors made a dataset of 83 K images consisting of 9 GANs and 4 DMs: half real and half synthetic [24]. There are three levels to the framework: authentic image vs. AI image, GAN vs. DM, and identifying specific architecture. This approach was over-accurate, a rate that surpassed previous methods [24]. The method is resilient to attacks, such as JPEG compression and resizing, making the method suitable for real applications. It also aids in forensic investigations by tracking the underlying source model of deepfakes. Multilevel classification is an effective method to improve detection performance.

3.3.4 Deep Convolution GAN

The authors proposed a model on a deep convolution network and GANs [41]. The planned model minimizes the loss of the discriminator and achieves more accuracy with each iteration. The proposed technique was implemented on celebA. The smaller batches were used, and the model provided 100% accuracy. The comparison GAN architecture used for deep fake detection is shown in Table 4.

3.4 Transformers

Transformers are neural network architecture for optimizing the processing of sequential data, per element order in each sequence, transformed by converting

Table 4 GAN's architecture accuracy comparison

Reference	Method	Dataset	Accuracy
Nadimpalli [22]	Embeds visible watermarking into GAN-generated fake images for proactive detection	FF++	88.5
Wang et al. [23]	Layer-wise neuron activation patterns	FF++	98.5
Guamera et al. [24]	Hierarchical multilevel classification framework	FF++	94
Preeti et al. [41]	Deep convolution GAN	CelebA	100

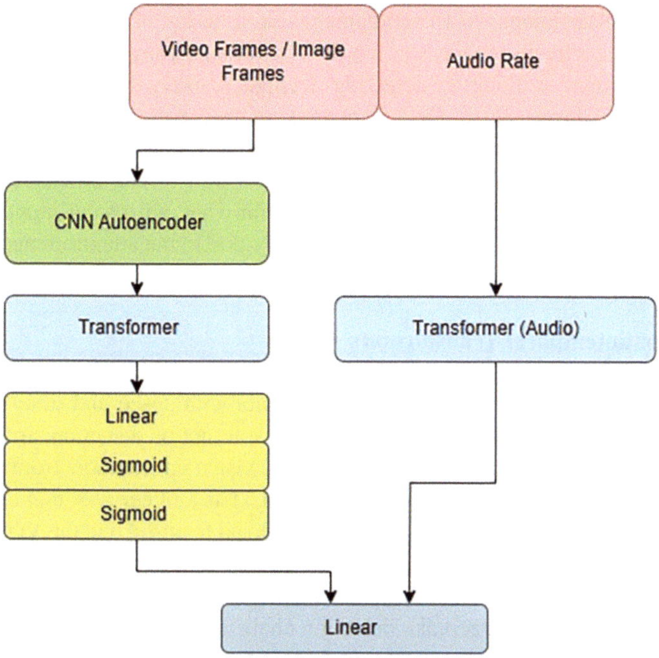

Fig. 6 Flowchart of transformer architecture

their interactions [25]. Through parallelization, transformers can learn relationships between distant points in a sequence simultaneously, something traditional models like RNNs struggle with. The central innovation of transformers is the self-attention mechanism. It allows the model to attend to different parts of the input sequence simultaneously. Further, it helps understand every element in relation to all the others, which creates a deep understanding of complex contexts. Also, they use multi-head attention to learn more independent ways to look at the data, in parallel, across different attention layers, hence richer representation. Adding positional information enables the model to learn order, which is needed for many tasks. Then, each element of that data is passed through a feed-forward (FF) neural network for each element. Transformers also apply techniques like layer normalization and residual connections to psychosocially ensure stability during training and to enable the use of very deep architectures [25]. They prevent challenges such as vanishing gradients, which frequently occur in deep-learning models. Figure 6 shows the transformer architecture.

3.4.1 Vision Transformers (ViTs)

ViTs are a deep learning architecture specifically designed for computer-vision tasks, applying transformers to image classification and object detection problems. Instead of looking at the image, ViTs treat images as a series of smaller patches, like how transformers treat sequences of words in natural language. Self-supervised ViT models show state of the art in detecting deepfakes, likely due to their training on large, unrelated datasets [10]. One of the two common ways to leverage pretrained ViTs would be to either use their pretrained layers (backbones) for extracting features from a few levels or partially fine-tune the top few transformer blocks. Fine-tuning tends to outperform pretraining and helps explain what part of the input space the model attends to while generating output, largely due to the attention mechanism.

3.4.2 Spatiotemporal Transformers

Spatiotemporal transformers are used within data with space and time frequency, which means they can easily help to analyze videos, action detection, and deepfake detection. These transformers effectively capture visual information from individual video frames as well as the timing of events. HCiT is another way that combines a CNN and a ViT. As for HCiT, the CNN catches local features, and the ViT adapts its self-attention mechanism to find global patterns and context [11]. This architecture yields high performance in identifying deepfakes, evidenced by tests on datasets such as FaceForensics++ , the deepfake detection challenge, and Celeb-DF, where HCiT outperformed CNN-only as well as pure-ViT approaches.

3.4.3 Contrastive Data-Efficient Transformers

Contrastive data-efficient transformers (CoDeiT) employ a novel attention mechanism for hierarchical representation of the frames combined with contrastive learning to spot deepfakes in videos. It employs a spatiotemporal transformer to process video frames and feed the dependencies and context for long term. CoDeiT is able to capture complex features in videos, which captured the highest scores on deepfake detection tests on human labels compared with other models. To further boost the ability of discriminating diverse features learned by CoDeiT [12], it utilizes a contrastive learning strategy with InfoNCE loss. The model is able to achieve superior performance in the task of differentiating between real and fake items using the methodology presented. Approaches using detailed data augmentation are applied as a model quality checker for improved performance across several datasets [12]. Their unique standard architecture is utilized to learn spatial–temporal patterns, and through the use of attention mechanism techniques, they apply contrastive learning for deepfake detection.

3.4.4 Convolutional Vision Transformer

The convolutional ViT was employed for the detection of deepfakes [42]. The model uses two components, that is, CNN and ViT. Learnable features were extracted

Table 5 Transformer architecture accuracy comparison

Author and reference	Model	Dataset	Accuracy
Khormali and Yuhan [10]	Vision transformers	FF++	99.41
Kaddar et al. [11]	Spatiotemporal transformers HCiT	FF++	96
Zakkam et al. [12]	Contrastive data efficient transformers XL	Deepfake Detection Challenge (DFDC)	86.9
Wodajo and Atnafu [42]	Convolutional vision transformer	Deepfake Detection Challenge (DFDC)	91.5

through CNN, and further, ViT uses these learnable features to categorize them through the attention mechanism. The technique was trained on the deepfake detection challenge dataset, attaining 91.5% accuracy. The transformer architecture for deepfake detection is given in Table 5.

4 Results and Discussion

This section discusses the results of the different methodologies for deepfake detection. Figure 7 depicts the accuracy of the different models used for deepfake detection. It has been observed that GANs have the highest accuracy of 95.25%, followed by LSTM and hybrid that gives 93.76%. The CNN model provides an accuracy of 93.68%. The transformer model gives an accuracy of 93.45%. The DNN has different application areas. These different models are applied to diverse areas. DNNs are a class of artificial neural networks characterized by multiple layers of interconnected neurons that aim to model complex patterns in data [15]. By leveraging stacked layers, DNNs can learn hierarchical representations, from low-level features (e.g., edges and textures) to high-level abstractions (e.g., shapes and object categories) [16]. Their flexibility and scalability make them a powerful tool for a variety of tasks, including image classification, natural language processing, and anomaly detection. Transformers, such as ViT and swin transformers, excel at capturing global relationships across images and video frames, identifying fine-grained and context-aware manipulations [17]. DNNs also analyze temporal inconsistencies in video data through RNNs and LSTM networks, detecting anomalies in motion patterns and unnatural transitions. Moreover, DNNs identify subtle artifacts introduced by manipulation techniques, such as pixel-level irregularities and boundary artifacts, with hybrid architectures like CNN-LSTM combining spatial and temporal analysis for enhanced detection [18].

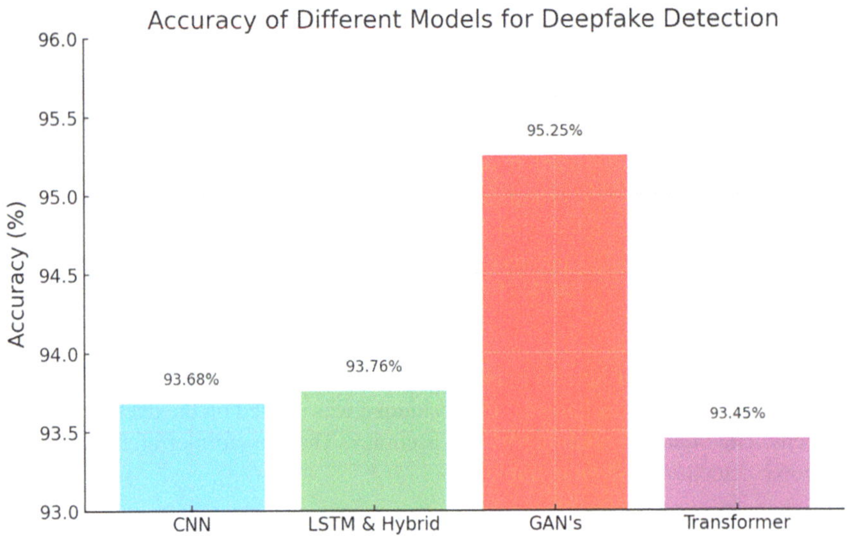

Fig. 7 Depicting the accuracy of different models used for deepfake detection

Pretrained models like InceptionResNetV2 demonstrate strong generalization, effectively adapting to various deepfake methods, including unseen techniques. Multimodal approaches further extend DNN capabilities, enabling the detection of mismatches in audio-visual data, such as lip-sync discrepancies [19]. Additionally, lightweight DNN architectures support real-time detection, making them ideal for live-streaming platforms and social media monitoring.

Deep learning models have a diverse area of application. Figure 8 depicts the different application areas of deep learning models. It includes computer vision, natural language processing, healthcare, fraud detection, and many other areas. The deep learning models are used for object detection [28], image recognition [30], facial recognition [27], self-driving cars [29], etc. It also has its applications in machine translation for translating text from one language to another, virtual assistant for creating intelligent conversational agents [31], sentiment analysis [32], spam filtering [33], text summarization [34], etc. It also applications in disease detection [36, 37], drug discovery [38], patient monitoring [35], personalized medication, etc. The models can be helpful in data analysis, cybersecurity, and recommender systems [39, 40].

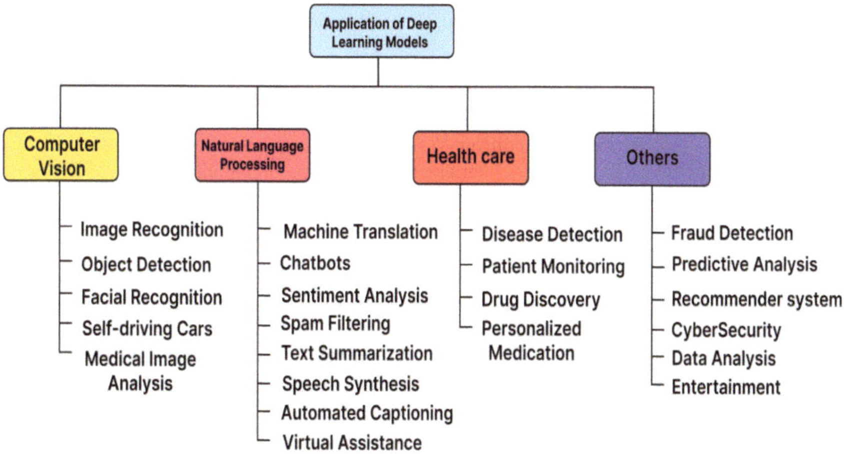

Fig. 8 Applications of deep learning models

5 Conclusion

This study presents various deep learning architectures—CNNs, LSTMs, GANs, and transformers for deepfake detection, focusing on their ability to analyze spatial and temporal consistencies. Transformers emerged as the most effective, excelling in capturing global spatial and temporal relationships through self-attention mechanisms, with ViTs achieving 99.41% accuracy on FF++ . CNNs, like XceptionNet and EfficientNet, demonstrated strong spatial feature extraction, achieving up to 98.75% accuracy on Celeb-DF, but struggled with temporal modeling. Hybrid CNN-LSTM models combined spatial and temporal analysis, leveraging optical flow features to detect unnatural motion patterns, achieving 91.21% accuracy on FF++ , though they lagged in efficiency. GAN-based methods, such as FakeSpotter, showed robustness with 98.5% accuracy by analyzing neuron activation patterns but faced computational challenges. It has also been observed from the results that overall GANs have the highest accuracyfollowed by LSTM and hybrid, CNN, and transformer. It has been observed that there are different parameters that need to be considered while evaluating the accuracy of a model such as facial image, expression, and size of dataset.

The significance of this study lies in its comprehensive evaluation of state-of-the-art models, providing a roadmap for future research in deepfake detection. By highlighting the strengths of transformers in handling both spatial and temporal inconsistencies, this study underscores their potential for real-time applications, such as social media monitoring and live-streaming platforms. Future work can focus on optimizing these models for scalability, integrating multimodal approaches (e.g., audio-visual analysis), and enhancing robustness against adversarial attacks.

References

1. Jones VA (2020) Artificial intelligence enabled deepfake technology: the emergence of a new threat (Master's thesis, Utica College)
2. Kalemullah AM, Prakash P, Sakthivel V (2024) Deepfake classification for human faces using custom CNN. In: 2024 7th International conference on circuit power and computing technologies (ICCPCT). IEEE, vol 1, pp 744–750
3. Liu Y, Zhang Y, Wang Y, Hou F, Yuan J, Tian J, He Z (2023) A survey of visual transformers. IEEE Trans Neural Netw Learn Syst
4. Shad HS, Rizvee MM, Roza NT, Hoq SA, Monirujjaman Khan M, Singh A, Bourouis S (2021) [Retracted] comparative analysis of deepfake image detection methwod using convolutional neural network. Comput Intell Neurosci 2021(1):3111676
5. Afchar D, Nozick V, Yamagishi J, Echizen I (2018) Mesonet: a compact facial video forgery detection network. In: 2018 IEEE International workshop on information forensics and security (WIFS). IEEE, pp 1–7
6. Yasser B, Hani J, El-Gayar S, Amgad O, Ahmed N, Ebied HM, Salah M (2023) Deepfake detection using efficientnet and xceptionnet. In: 2023 Eleventh international conference on intelligent computing and information systems (ICICIS). IEEE, pp 598–603
7. Mitra A, Mohanty SP, Corcoran P, Kougianos E (2020) A novel machine learning based method for deepfake video detection in social media. In: 2020 IEEE international symposium on smart electronic systems (iSES)(Formerly iNiS). IEEE, pp 91–96
8. Yadav P, Jaswal I, Maravi J, Choudhary V, Khanna G (2021) DeepFake detection using inceptionResNetV2 and LSTM. In: international conference on emerging technologies: AI, IoT, and CPS for science technology applications
9. Al-Dulaimi OAHH, Kurnaz S (2024) A hybrid CNN-LSTM approach for precision deepfake image detection based on transfer learning. Electronics 13(9):1662
10. Khormali A, Yuan JS (2022) DFDT: an end-to-end deepfake detection framework using vision transformer. Appl Sci 12(6):2953
11. Kaddar B, Fezza SA, Akhtar Z, Hamidouche W, Hadid A, Serra-Sagristà J (2024) Deepfake detection using spatiotemporal transformer. ACM Trans Multimed Comput, Commun Appl
12. Zakkam J, Jayaraman U, Sahayam S, Rattani A (2025) Codeit: contrastive data-efficient transformers for deepfake detection. In: international conference on pattern recognition. Springer, Cham, pp 62–77
13. Namatēvs I (2017) Deep convolutional neural networks: structure, feature extraction and training. Inf Technol Manag Sci 20(1):40–47
14. Omer ZT, Abbas AH (2022) Image anomalies detection using transfer learning of ResNet-50 convolutional neural network. Indonesian J Electr Eng Comput Sci 27(1):198–205
15. Samek W, Montavon G, Lapuschkin S, Anders CJ, Müller KR (2021) Explaining deep neural networks and beyond: a review of methods and applications. Proc IEEE 109(3):247–278
16. Bouwmans T, Javed S, Sultana M, Jung SK (2019) Deep neural network concepts for background subtraction: a systematic review and comparative evaluation. Neural Netw 117:8–66
17. Liu Z, Lin Y, Cao Y, Hu H, Wei Y, Zhang Z, Guo B (2021) Swin transformer: hierarchical vision transformer using shifted windows. In: Proceedings of the IEEE/CVF international conference on computer vision, pp 10012–10022
18. Liang P, Liu G, Xiong Z, Fan H, Zhu H, Zhang X (2023) A facial geometry based detection model for face manipulation using CNN-LSTM architecture. Inf Sci 633:370–383
19. Nawaz M, Javed A, Irtaza A (2024) A deep learning model for faceSwap and face-reenactment deepfakes detection. Appl Soft Comput 111854
20. Paul OA (2021) Deepfakes generated by generative adversarial networks
21. Dash A, Ye J, Wang G (2023) A review of generative adversarial networks (GANs) and its applications in a wide variety of disciplines: from medical to remote sensing. IEEE Access 12:18330–18357

22. Nadimpalli AV, Rattani A (2024) Proactive deepfake detection using gan-based visible watermarking. ACM Trans Multimed Comput Commun Appl 20(11):1–27
23. Wang R, Juefei-Xu F, Ma L, Xie X, Huang Y, Wang J, Liu Y (2019) Fakespotter: a simple yet robust baseline for spotting ai-synthesized fake faces. arXiv:1909.06122
24. Guarnera L, Giudice O, Battiato S (2024) Mastering deepfake detection: a cutting-edge approach to distinguish GAN and diffusion-model images. ACM Trans Multimed Comput Commun Appl 20(11):1–24
25. Chitty-Venkata KT, Emani M, Vishwanath V, Somani AK (2022) Neural architecture search for transformers: a survey. IEEE Access 10:108374–108412
26. Srinivas V, Angara, M, Chamarthi S, Gangisetti, S, Lingala V (2024) Deepfake detection based on temporal analysis of facial dynamics using LSTM and ResNeXt architectures. J Image Process Intell Remote Sens 4:47–54. https://doi.org/10.55529/jipirs.43.47.54
27. Sun X, Wu P, Hoi SC (2018) Face detection using deep learning: An improved faster RCNN approach. Neurocomputing 299:42–50
28. Wu X, Sahoo D, Hoi SC (2020) Recent advances in deep learning for object detection. Neurocomputing 396:39–64
29. Fujiyoshi H, Hirakawa T, Yamashita T (2019) Deep learning-based image recognition for autonomous driving. IATSS research 43(4):244–252
30. Islam MT, Siddique BNK, Rahman S, Jabid T (2018) Image recognition with deep learning. In: 2018 International conference on intelligent informatics and biomedical sciences (ICIIBMS). IEEE, vol 3, pp 106–110
31. Vishnu R, Krishna Prakash N (2022) Mobile application-based virtual assistant using deep learning. In: Soft computing and signal processing: proceedings of 3rd ICSCSP 2020. Springer Singapore, vol 2, pp 609–617
32. Tang D, Zhang M (2018) Deep learning in sentiment analysis. Deep learning in natural language processing 219–253
33. Roy PK, Singh JP, Banerjee S (2020) Deep learning to filter SMS Spam. Futur Gener Comput Syst 102:524–533
34. Yousefi-Azar M, Hamey L (2017) Text summarization using unsupervised deep learning. Expert Syst Appl 68:93–105
35. Rajan Jeyaraj P, Nadar ERS (2022) Smart-monitor: Patient monitoring system for IoT-based healthcare system using deep learning. IETE J Res 68(2):1435–1442
36. Ferentinos KP (2018) Deep learning models for plant disease detection and diagnosis. Comput Electron Agric 145:311–318
37. Chen YW, Jain LC (2020) Deep learning in healthcare. Paradigms and Applications; Springer: Berlin/Heidelberg,Germany
38. Liu M, Li C, Chen R, Cao D, Zeng X (2024) Geometric deep learning for drug discovery. Expert Syst Appl 240:122498
39. Steck H, Baltrunas L, Elahi E, Liang D, Raimond Y, Basilico J (2021) Deep learning for recommender systems: a Netflix case study. AI Mag 42(3):7–18
40. Ghillani D (2022) Deep learning and artificial intelligence framework to improve the cyber security. Authorea Preprints
41. Kumar M, Sharma HK (2023) A GAN-based model of deepfake detection in social media. Procedia Computer Science 218:2153–2162
42. Wodajo D, Atnafu S (2021). Deepfake video detection using convolutional vision transformer. arXiv:2102.11126

DeepFex: Explainable and Secure Deepfake Detection System

R. Karthick Manoj and S. Aasha Nandhini

1 Introduction

Modern digital media develops artificial intelligence technologies combined with deep learning features rapidly. The numerous uses of deepfake technologies raise serious questions as they supply exact reproductions of actual human characteristics from seem-to-speak and physical movement, so permitting deepfakes generation. Although deepfakes provide creative possibilities and entertainment value, they also bring ethical dilemmas, as well as social challenges and safety issues [1]. Synthetic products cause customers to lose faith in digital data, which is already declining. Deepfakes research is still highly important for new approaches to discovery as well as for constructing anti-manipulation protection and disinformation defensive mechanisms. DeepFex's user-driven synthetic video content identification technology presents a unique approach for deepfake detection. DeepFex creates scalable solutions for robust privacy-enabled detection of modified media assets by means of blockchain-based and federated learning (FL) integration [2]. The study looks at DeepFex development schedules, including technical integrations, to examine system implications on modern digital safety requirements.

R. Karthick Manoj (✉)
Department of Electrical and Electronics Engineering, AMET Deemed to Be University, Kanathur, India
e-mail: Karthickmanoj.r@gmail.com

S. Aasha Nandhini
Department of Electronics and Communication Engineering, Sri Sivasubramaniya Nadar College of Engineering, Kalavakkam, India

A. Kumar et al. (eds.), *Blockchain and Federated Learning Synergy for Privacy-Focused DeepFex Solutions*, Blockchain Technologies,
https://doi.org/10.1007/978-981-95-1394-9_4

1.1 The Rise of Deepfakes

The discovery of deepfakes occurred two years ago, leading to the development of sophisticated versions, which increased production speed [3]. Early simulation technologies emerged from research conducted at AI firms and open-source software projects, and these technologies now fool most human observers into believing them real. Generative adversarial networks (GANs) merged with autoencoders and transformers enabled developers to produce enhanced deepfakes capable of real-time generation of speech audio alongside facial emotional expressions and body movements [4]. New tools became easier to obtain across various platforms, leading to their heightened adoption throughout social media networks, while entertainment platforms, alongside problematic guessing events of political deception and instances of cyberbullying and identity theft and additional criminal conduct. General understanding of deepfake data dissemination has sparked public issues concerning institutional and digital business remedies toward the problem, jointly with public organizations. Deepfake creators keep ahead in development, trying to outcompete those working on detection systems. Artificial intelligence keeps developing its creative ability, but society depends mostly on dependable, scalable detection systems that should always have an edge over expanding threats [5].

1.2 Threat Landscape in the Digital Era

Digital technology turned knowledge generation and information distribution, along with learning access, into new national approaches to link individuals and provide unlimited access to data [6]. Thanks to the development of deepfakes technologies, the security area is developing rapidly. Deepfakes become a useful tool for manipulation because of the fast false information in our linked society. Deepfakes are used in many different fields to perform political manipulations, acquire business intelligence, create social designs, and propagate slander, generating different worldwide security issues. Deepfakes artificially erode media validity, therefore upsetting democratic institutions and causing great anguish to victims when their material is altered. The application of deepfake technology by government agencies alongside nongovernmental entities for unapproved purposes creates risks to national security systems [7]. Advanced deepfake technology became available through its democratic process, allowing enemy operatives to carry out clandestine operations in any location. Digital systems verification methods fail when synthetic media appears natural and realistic because these traditional verification techniques become obsolete. To preserve the integrity of digital systems, digital material preservation calls for both technological advancement and a combined multidisciplinary effort between government agencies and public awareness campaigns.

1.3 Need for Privacy-Preserving Detection

Deepfake detection system training and operating data protection needs more than technological solutions. Regular deep learning detector training depends on centralized data storage systems while requiring large amounts of sensitive data to operate. Data centralizing system users face concerns about data breaches as well as the loss of ownership and potential illegal espionage activities [8]. Emphatically, we need newly developed privacy-protecting detection techniques because the complex nature of biometric data security makes it impossible to make sensitive components such as voice recordings and face scans unrecognizable. Using blockchain technology and deepfake detection methods allows systems to find deepfakes securely with an unalterable open network design. The Federation training procedure enables distant devices to establish cooperative models through data-free processing. The model-building process maintains anonymity through simultaneous model development on numerous devices combined with multiple participants. A secure deployment model for deepfakes arises through the combination of tamper-proof blockchain technology with this method. The privacy-protecting solutions offered by DeepFex architecture lead to revolutionary changes in digital authentication methods while safeguarding privacy measures for deepfake detection.

2 Understanding Deepfakes

2.1 Definition and Types of Deepfakes

The term "deepfake" defines synthetic media that replaces a person depicted in original imagery with a different person's appearance [9]. The creation of hyperrealistic forgeries uses deep learning techniques with GANs to oppose two neural networks (the generator and discriminator) to generate more authentic material. Covering a wide range of content, video, audio, and images, the word "deepfake" combines "deep learning" with "fake [10]." Several types of deepfake technology have recently become available with specific processes for achieving their effects. Often masterfully blending emotions and gestures, face-swapping deepfakes replace the face of a subject in a video. Lip-sync deepfakes basically match the lip gestures of a person in a video to match other audio information, therefore making the subject look to be uttering different words [11]. Voice cloning is another well-known field in which, sometimes, with minimal input data, the audio of a person's speech is electronically synthesized to imitate their voice. More sophisticated full-body deepfakes are also using person's whole body movements and postures as inspiration. These numerous forms draw attention to the flexibility and probable danger of deepfakes, which may produce rather convincing and false digital content on several platforms.

2.2 Techniques Used to Generate Deepfakes

Complex machine learning systems with deep neural networks create most deepfakes. Most deepfake generating systems rely on GANs that comprise two opposing models that perform data creation and data validity assessment [12]. Through an adversarial process, the generator develops progressively realistic outputs until the discriminator stops identifying original from artificial data. The method of using autoencoders enables face swapping operations through an encoder compression process, which produces latents, and then the decoder reconstructs images for target domains. Two recent advancements in deepfake technology include diffusion models along with transformer-based models, which excel at creating realistic images and videos. Deepfake creation became easier through transfer learning and few-shot learning since they decreased model training demands and data requirements [13]. Deepfakes maintain visual accuracy through facial landmark detection, together with posture estimate and temporal coherence modeling techniques to keep their content consistent throughout time. The advancement of technical capabilities has reduced the requirements for resources during sophisticated deepfake creation, which expands both the harmful and beneficial usage potentials.

2.3 Real-World Examples and Impact

The existence of deepfake technology has resulted in numerous global events that have had severe societal and political repercussions, as well as economic losses. The earliest and most well-known use of deepfake technology was illegal celebrity appearances for pornographic adult film performers. This concerning tendency enabled the identification of regular occurrences of personal privacy invasions and psychological harm done to people. Deepfakes have such bad consequences as disseminating misleading information while mimicking political leaders worldwide. False film footage depicting politicians making controversial comments on online sites that undermined public trust and may have changed the result of elections. Deepfakes disseminating modified news content across social media platforms aggravate societal conflicts [14]. Deepfaked content raises issues around permitted content legitimacy, even if it improves visual effects and dubbing for entertainment purposes. Deepfakes challenge human ability to detect actual material from synthetic information, so society dismisses real items as fake and causing psychological stress. The collection of events highlights the importance of building robust detection systems in parallel with legislation aimed at reducing deepfake technology risks.

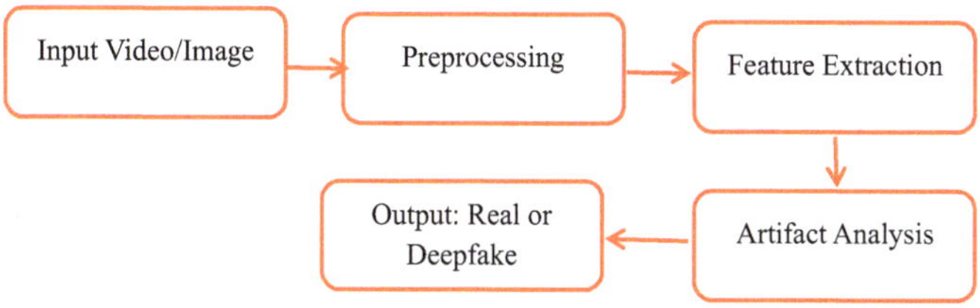

Fig. 1 General pipeline for deepfake detection in visual media

2.4 Deepfake Detection Techniques

2.4.1 Traditional Detection Approaches

When deepfake detection protocols were developed, the deepfake production techniques had basic qualities that made artificial traits obvious through visual indications and auditory inconsistencies [15]. These detection tactics examine errors that happen in regulated material-based processing through manual observations along with statistical quantification. Initial deepfakes contained multiple noticeable artificial defects such as synthetic eye movements combined with facial piece-backdrop mismatch and unnatural facial behavior. The failure to match real and fake content exists because the generative model lacks proper training or the rendering technology functions insufficiently. Multiple image forensics methods applied to this field involve the analysis of head posture alignment and compression artifacts, and noise pattern recognition. These methods function at low cost while needing small datasets and absent neural network processing, hence they work for lightweight applications. The methods prove less effective against current deepfakes developed through powerful GANs that generate data which appears virtually indistinguishable from real content [12]. Traditional detection methods demonstrate limitations when it comes to flexibility since they fail to adjust to emerging manipulation technologies and lack the ability to handle different media types. Traditional methods established fundamental principles that help advanced detection systems pair machine learning models with handmade detection elements.

Figure 1 visualizes how most deepfake detection systems operate through sequential steps, which start from input acquisition and end with classification. The detection process consists of four steps, including preprocessing, feature extraction, artifact analysis, and announces whether media content is real or manipulated.

2.4.2 Deep Learning-Based Detection

The illustration presents deep learning models for detecting deepfake content. The analytical process introduces input data as visuals or video through deep neural networks that extract features using CNNs or RNNs before moving to the classification stage. The learned features proceed to the classification layer, which

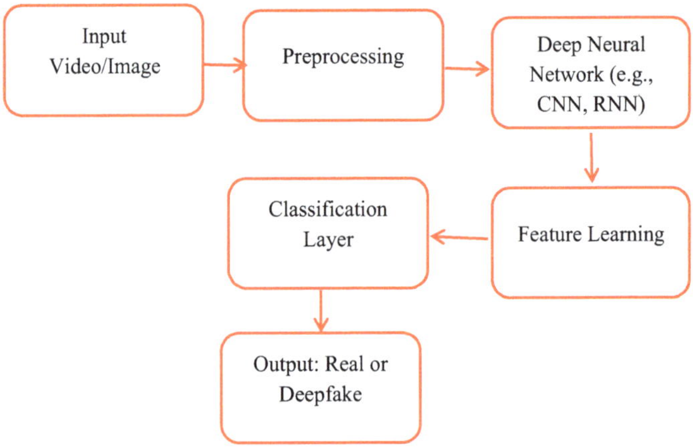

Fig. 2 Deep learning-based pipeline for deepfake detection

gives an output verdict about authentic or synthetic content. The advent of deep learning technology established a new era for deepfake detection through learning processes that analyze unprocessed data directly. CNNs allow extensive identification of subtle abnormalities found in images and video frames. The models operate below human perception levels to identify subtle face anatomical irregularities and match or mismatch color patterns and surface textures. Although successful training of CNN models yields improved accuracy and generalization abilities, the networks require extensive datasets with annotations for their development. LSTM networks derived from RNNs show mastery in analyzing temporal dependencies of video frames, which facilitates the investigation of transitioning movements connecting frames to detect artificial movements and lip-sync mismatches (Fig. 2).

Autoencoders are especially useful in anomaly detection situations to reconstruct original material before comparing it against input data for detecting possible modifications. The advent of capsule networks provides enhanced spatial hierarchical functionality and resistance against slight input alterations, which proves crucial for stopping adversarial attacks. Numerous trade-offs involving computing expenses and training complexity as well as interpretability emerged among these models. The aggregated strength of these models created significant improvements in the field that enable automated detection of complex deepfakes, which occur in real-world conditions.

2.4.3 Challenges in Detection

Deepfake detection faces various complex obstacles despite the advancement in conventional and deep learning-based methods [10]. Models that receive training for a single deepfake alteration or dataset tend to identify unknown modifications as different from their training domain. Social media environments that undergo substantial material quality and structural modifications face major problems with generality because of it. The creation of deepfakes by attackers specifically to avoid the latest detection technology creates an intensive additional adversarial threat. The attacks exploit neural network weaknesses to create undetectable shifts that lead models to generate wrong results. Two severe barriers limit both the development and acquisition of adequate training and assessment data: scarce data resources and prohibitive ethical requirements for obtaining enough labels.

The process of threat identification proves difficult for modern technology because it targets mobile devices and high-definition video feeds, among other equipment. Applications that include security forces along with journalism and law enforcement need low-latency, high-accuracy solutions to function in real time. Large neural network systems produce judgment outputs that remain difficult for consumers to understand, which leads to reduced trust in security-related sectors. The development of reliable detection systems necessitates cooperation between legal experts, cybersecurity experts, ethical counselors, and AI professionals.

2.4.4 Taxonomy of Detection Methods

Various researchers have developed different taxonomies to organize available deepfake detection methods by their functional principles while defining their targets and applicable fields of use. Most common detection taxonomies identify two major detection categories, which are passive and active approaches. The detection methods rely solely on examining provided materials to uncover visual and aural, and temporal domain inconsistencies [10]. They do not need additional information. Active detection functions embed small cryptographic signatures along with watermarks to allow future authentication testing. The particular deepfake models require photographic content, so media detection requirements determine how these models should be categorized.

Instrumental analysis establishes a classification system that divides detection systems into three categories, from emotional discrepancies at the semantic-level up to temporal discrepancies at the frame-level before reaching color distinctions at the pixel-level. The detection architecture functions as a separate approach to model classification using CNNs and RNNs with transformer-based hybrid models among them. Such detection systems help users choose suitable instruments for specific application requirements. Related research into deepfake detection models across different domains helps to bridge gaps in the existing knowledge base, and new models for analyzing combined audio-visual deepfake alterations function as a beneficial research outcome. Field-based innovation can be developed through well-organized classification systems that exist for researchers.

3 The DeepFex Framework

3.1 Overview of the DeepFex System

The DeepFex system represents a modern method for handling escalating challenges that stem from deepfake technology. DeepFex combines blockchain technology with FL and deep learning to establish a privacy-preserving detection platform for deepfakes, which improves its strength and continuous development. The network operates with distributed operations that connect personal devices to institutional servers through edge computing to establish proactive privacy-protected detection methods [16]. Through distributed FL, DeepFex achieves its main breakthrough by placing machine learning models directly on local datasets to allow distributed training. Safety goals relating to privacy law compliance can be achieved by keeping sensitive raw data stored locally and exchanging model parameters or gradients instead of the raw data. The methodology permits users to present data from various backgrounds and global locations, thus improving privacy measures and enabling diverse and inclusive training data sets. DeepFex functions effectively in both law enforcement agencies and newsrooms, together with social media platforms, thanks to distributed learning that allows real-time responsiveness and the scalability that improves its usefulness.

3.2 Key Components and Architecture

DeepFex architecture features a modular design structure that combines several connected components that deliver exceptional performance together with high resilience levels. The most basic component of DeepFex architecture enables the input interface to acquire photographs as well as audio and video assets for processing. Standardization of data values takes place after the preprocessing module through its normalization and standardization phase that serves to prepare features. Another pivotal engine component called feature extraction engine detects vital auditory and visual temporal signals of manipulation using deep learning models that incorporate CNNs together with RNNs. The input characteristics of DeepFex derive their elements from components within the FL core [17]. The core functions as a model updater and distributed training controller through differential privacy and secure multiparty computing aggregation techniques to allow global model improvement with protected individual data privacy. All modifications to models get stored on a distributed blockchain ledger through immutable transaction records. Smart contracts within the ledger execute multiple tasks, which include model check analysis and authenticity verification, and access rights administration functions. The reporting interface presents audit results to individuals who need them in order for the system to demonstrate accountability at all operational levels. System protection from rising security issues is ensured by the layered design, which provides independent upgrade capabilities to each component.

3.3 Integration with Privacy-Preserving Technologies

DeepFex distinguishes itself from other detection systems by means of several privacy-defensive strategies. For the healthcare sector, several companies utilize FL to teach models without user data. This system guarantees communication and safe data storage. In FL, distributed nodes preserve user data ownership while safeguarding national borders and satisfying worldwide privacy requirements throughout run-through computations. DeepFex protects privacy in its aggregation stage by using differential privacy. The mathematical method replicates aggregated data changes with noise, therefore hiding specific user information. Security multiparty computation (SMPC) lets users guard their input data identity, therefore enabling training participants to move safely with unknown peers [18]. Blockchain technology uses open lines of communication to link responsible systems utilizing various operating systems. Gatehouse contracts on blockchain systems control program administration utilizing distributed identity management under limited access. Blockchain gatehouses prevent access and use distributed identities for management. Its integration of blockchain offers audit trails to confirm discovery outcomes and removes privacy and confidence issues. Modern safe AI enables DeepFex to provide a digital era new solution combining performance analytics with tight privacy protection.

4 Role of Federated Learning in DeepFex

4.1 Fundamentals of Federated Learning

Separating data across server platforms allows users of FL to create shared models. With FedAvg, FL systems maximize model changes to provide desired results. Every device changes its local parameters during model training; the server improves the model by aggregating the values. The modern technologies restrict data to certain areas and protect it. FedAvg is one of the successes with optimization techniques used in FL models. The method should operate within the framework of data protection regulations such as HIPAA and GDPR to control data movement. FL is the core purpose of the DeepFex architecture as it guarantees that personal user data—such as facial images, voice recordings, or behavioral patterns remains on every device [19]. The learning exercise emphasizes data points instead of gathering information from user devices, therefore producing a self-contained and privacy-compliant solution. By means of the intentional placement of this concept in DeepFex's fundamental architectural core, the system removes privacy issues and provides access to anyone who so chooses, therefore transcending geographical limitations or data privacy criteria.

4.2 Training Deepfake Detection Models on Edge Devices

Among other things, DeepFex framework with FL capability stands out for its ability to natively train deepfake detection models on edge devices. Edge computing server systems run from the original data source, unlike being dependent on far-off cloud-based systems. Local computers, security cameras, and cellphones all run this way. DeepFex helps devices to evaluate locally kept data or information gathered from their surroundings, therefore improving and preserving their detection systems. Together with compressed ResNets, the system uses MobileNets as deep neural network versions that retain great performance with low accuracy loss. Algorithmic signals that find differences between audio and video data and aberrant face changes with poor eye motions can help to discover deepfake defects in inputs. Before reaching either servers or coordinating nodes, all new model analyses are encrypted. Under safe multiparty computing SMPC homomorphic encryption and differential privacy are used in data security techniques to prevent information exposure while processing data. The DeepFex system detects distant nodes using this framework, therefore enhancing the performance accuracy of the global generalizable model. Deepfake detectors incorporated into edge devices that operate outside of continuous internet connectivity, as well as reducing data demands, allow security operations alongside media events to become practical.

4.3 Benefits: Data Sovereignty and Privacy

The combination of enhanced privacy standards and ownership control enables FL methods to reach DeepFex standards, thus producing multiple useful results. Data sovereignty provides ownership control that establishes storage locations and operating procedures for personal data across generic individual and company entities [20]. Users of contemporary centralized AI systems must store their data on distant servers, which results in multiple security issues, including privacy violations together with illegal data usage, and data breaches. Total management of data provides absolute power for users to safeguard their private messages, court documents, medical files, and important data, which could result in catastrophic consequences if exposed. The ability to maintain complete oversight ensures users retain full authority over their private communications, judicial documents, medical records, and sensitive information, which would result in catastrophic damage if compromised. The targetability of distributed learning data makes it harder for hackers to focus on centralized data repositories.

The DeepFex system uses distributed training across multiple systems to protect network nodes from attackers who target their data storage resources. Mathematical underpinnings of differential privacy provide secure dissemination of updates free from any compromise to personally identifiable data. If deepfake detection technology is to be adopted globally, it must include components that inspire user confidence and acquire their permission. DeepFex improves deepfake detection performance, thereby resolving privacy issues; yet, FL helps to protect user autonomy.

5 Blockchain Integration for Trust and Traceability

5.1 Blockchain for Data Integrity and Auditability

DeepFex's design mostly relies on blockchain technology, as blockchain provides a distributed ledger guaranteeing unchangeable system data and complete auditing powers. Deepfake detection and digital forensics analysis still depend mostly on building real evidence of media origin and provenance. The detection logs coupled with model updates and data sources must be consistent, as regulated content can readily enter digital ecosystems in a fast-paced setting. Forms component of blockchain architecture, producing a tamper-resistant system solution to solve the problem. Starting with model changes to detection results and access requests, which connect back to previous blocks, the blockchain system employs cryptographic techniques to record every form of transaction or event. The architecture of the blockchain guarantees that changes stay absolutely observable and trackable as well as unchangeable and verified. The DeepFex solution enhances accountability by including blockchain records that allow independent verification from consumers and authorities [21]. When the system finds such media files, the blockchain keeps detection metadata for deepfake files together with timestamps, device IDs, and model versions. Tracking who created limited content lets reporters cross-check claims without depending entirely on source material. Since this protection is essential to keep public confidence in deepfake detection technology, the distributed blockchain architecture removes security threats from single points of contact and authority.

5.2 Smart Contracts for Access Control and Provenance

Smart contracts are self-executing software combined into a blockchain that automatically applies user-defined criteria. DeepFex smart contracts are part of the system to offer data and model update provenance verification, access management, and compliance assurance. Deepfake detection's privacy-sensitive feature calls for users to maintain specific knowledge-sharing access limits from particular stakeholders.

Raw metadata is needed by the law enforcement agency; journalists or public users simply need to confirm the veracity of the proof. Smart contracts can manage and create roles by means of access privileges executed in line with predefined rules. Before the model can be included in a block, the blockchain needs a smart contract to confirm an edge device model update through identity checks.

Smart contracts users authenticate training data as well as the source records of detection outcomes. Verification that training or assessment data is unaltered helps to guarantee the necessity of ethical data source standards. Through rapid process automation as well as low latencies, automated verification using smart contracts both lessens reliance on middlemen and improves system dependability. Through its auditability technology, every access and action on the blockchain ledger becomes traceable and offers an official, verifiable record fit for academic repeatability as well as dispute resolution and compliance needs. Smart contracts are necessary for the DeepFex system to fully incorporate blockchain and guarantee enforceability by means of their technological logic application.

5.3 Synergy with Federated Learning in DeepFex

DeepFex builds security and trust via FL and blockchain. FL distributes training across devices to guarantee privacy, secure model updates, and validation coordination are thus required [22]. Blockchain technologies satisfy this demand with their validation mechanisms. Following local instruction, an edge device delivers a message updating a blockchain. Using consensus protocols and smart contract rules, the blockchain network checks cryptographic signatures before distributed validators approve or reject model modifications. Once approved, the allowed change comes under the global model aggregate. Constructing the final model helps distributed validation remove unethical input and central power.

Blockchain systematically records real-time participant input while maintaining full transparency and accountability in the system. Security tracks for audit history remain indispensable for regions prone to high-risk situations. DeepFex implements two-layer protection through blockchain's immutable technology and FLs secure privacy models. The combination of local computation with obfuscated updates enables secure protection of privacy in tamper-resistant and openly accessible records, which enhances trustworthiness. Partnerships enable the deployment of novel incentive methods. Users who submit data or model enhancements for blockchain incentive systems earn reputation points combined with blockchain payment rewards. User participation grows faster when they interact with models through these techniques. Blockchain systems integrated with FL develop an ethical structure that enables technical advancement while maintaining social acceptance. DeepFex stands as a revolutionary solution that fights deepfakes and digital disinformation.

6 Performance Evaluation and Benchmarks

6.1 Datasets Used for Training and Testing

The framework validity of DeepFex depends heavily on selecting proper benchmark datasets for deepfake detection model training and testing purposes. DeepFex ensures potent and scalable detection capabilities through its usage of public and privately sourced datasets, which cover video images and audio formats. The most prevalent datasets for deepfake detection include FaceForensics together with deep fake detection challenge (DFDC), Celeb-DF, and DeeperForensics-1.0.0. Each database provides unique prospects alongside specific challenges for users. The underdeveloped environment benefits from FaceForensics because it provides a broad selection of modified films across different compression configurations.

Released by Facebook AI, the DFDC dataset features a range of deepfake films created using several techniques and actors, therefore enabling an evaluation of the generalizability of the model. Celeb-DF is unique even for advanced detection models because of its excellent changes that quite mimic real material. DeeperForensics-1.0 also replicates real-world situations, including differences in occlusion, illumination, and resolution. FakeAV and ASVspoof abound in the databases. Celebrities have been trained to spot cross-modal anomalies in audio detection and voice copying. DeepFex uses virtual nodes to replicate distributed data environments, therefore simulating real-world deployment conditions and enabling FL. This varied dataset approach guarantees that the DeepFex framework is not only able to find current deepfake strategies but also resistant to future and newly developing manipulation methods.

6.2 Accuracy, Latency, and Privacy Trade-Offs

The performance of DeepFex framework calls for an examination of all three important factors between accuracy, latency, and privacy issues. To project accuracy levels, the conventional classification criteria include precision, recall, F1-score, and area under the ROC curve (AUC). DeepFex performs well in many studies since, evaluated on FaceForensics + + combined with Celeb-DF datasets, it shows detection accuracy greater than 95%. The deployment procedure requires components beyond accuracy if it is to run effectively. Low-latency mobile applications and real-time systems depend on accurate media sample assessment times. Using model optimization strategies that trim and quantize delay, standard edge devices can perform DeepFex at sub-second rates.

Privacy identification calls for the comparison of subjective information exposure risk assessments with formal privacy commitments from several protection mechanisms [23]. While severe obfuscation techniques diminish accuracy, strong model compression lowers reaction time. Any approach for model compression-privacy protection has to sacrifice accuracy. Through resource-aware parameters as well as adaptive thresholds for threshold optimization, DeepFex allows deployment situation-specific performance. Although consumer apps will give quicker processing top priority, high-security settings will get privacy and auditability as their main considerations. DeepFex's adaptability helps to maintain necessary performance standards, so it achieves perfect usability throughout a wide spectrum of applications.

6.3 Comparative Analysis with Centralized Models

DeepFex validated its federated blockchain-based approach against conventional central deepfake detection techniques. Under optimal conditions, centralized models—which save all training data in a single location—have greater convergence rates and accuracy. DeepFex has several negative aspects like legal data entry issues, insufficient execution of several datasets, and unscalability. During comparative studies, the DeepFex model attained identical accuracy rates to state-of-the-art centralized models within 1–2% accuracy margin. Better generalization to fresh data made possible by the heterogeneous and dispersed training data within DeepFex allowed both system longevity and overfitting to be lowered concurrently.

DeepFex's use of edge processing devices removed data transmission and centralized processing needs' limits via parallel computing, therefore proving better scalability than centralized solutions. Blockchain technology, which provides an unchangeable log of every detective activity lacking in conventional systems, distinguishes DeepFex. Through which the system emerges as a technology breakthrough in deepfake detection development and acts as an organizational paradigm change, the research contrasts DeepFex with alternative approaches. DeepFex defines progressive improvements in the area by offering improved traceability, ethical compliance requirements, and similar or better performance than centralized services, therefore offering increased security procedures.

6.4 Challenges and Limitations

DeepFex has technological and operational challenges, even with privacy-protected deepfake identification. Another barrier is the limited processing capabilities of primary edge devices. Battery life, memory, and CPU power are all limited in smartphones and IoT sensors. Local execution of sophisticated deep learning models decreases performance or causes overheating by stressing these resources. Information distillation, quantization, and model pruning all assist in reducing model size, but they frequently provide poor accuracy. Background system management for continuous global model modifications might jeopardize system stability and user experience on low-resource devices. DeepFex's continuous deployment encourages users to establish a balance between model performance and hardware capacity.

Because semi-trustworthy participants are used in FL training systems, malicious nodes may introduce incorrect or biased data into them. Model degradation and false detection are now conceivable. To locate attacks, the assailant trains their local model with new data, then aggregates those modifications into one final attack. Unstable execution settings and unclean data sources that elude anomaly detection systems compromise model integrity. This is addressed in research using Byzantine fault tolerance, safe aggregation, and anomaly-based filtering; DeepFex does not employ these at scale, retaining speed and privacy.

Finally, scalability and expense are the key disadvantages of blockchains. Even in transparent, irreversible systems, blockchain causes delays and consumes resources. Because of their slow transaction speed and high cost, public blockchains are not suitable for recording model changes or detection findings. Even with advanced blockchains like private or permissioned chains, consensus, node synchronization, and smart contract execution all consume a significant amount of CPU. As the system approaches six digits, performance limitations emerge. Incorporating blockchain into FL systems necessitates the development of an appropriate architectural framework that balances efficiency, usability, and decentralization. Zero-knowledge proofs, off-chain storage, and lightweight consensus systems improve solutions, but they also add complexity that must be managed. DeepFex's deployment as a deepfake detection tool necessitates careful consideration of edge computing limits, FL security issues, and blockchain operating costs. In many applications, the framework provides security and scalability but operational solutions are required to maintain it viable.

6.5 Explainable Blockchain-Federated Learning Framework for Detecting Deepfake Manipulation in Wildlife and Veterinary Media

Deepfake technology's spread across human objectives has exacerbated the challenges associated with biodiversity studies, wildlife protection, and veterinary diagnoses. Manipulated animal photos and video feeds perplex conservation AI systems; therefore, medical records may be changed during gaps in ecological surveillance systems. Without sacrificing data security or system openness, an explainable FL architecture leveraging blockchain technology detects and visualizes deepfake changes in animal-based media. Blockchain-based code forms this framework. This suggested system operates using distributed devices belonging to veterinary clinics together with wildlife field cameras and research facilities, which do not send raw animal data due to ethical restraints, legal requirements, and proprietary limits. The system uses Grad-CAM for visual recognition of picture changes and an explainable mechanism based on SegCaps-CNN architecture with modifications. Using a FL approach that certifies the changes on a distributed blockchain platform, several users mix their network modifications of the model. By means of smart contracts, users may track tamper-resistantly alongside model deliberation management and execute auditable proprietary data exchanges amongst personalities.

Synthetic datasets of animal images and altered veterinarian scans were used to demonstrate the system's performance. The results showed that the visual output of Grad-CAM closely matched expert-annotated changed areas, so detection accuracy above 94%. Moreover, blockchain integration guaranteed traceability of all changes without revealing private animal data and avoided model poisoning. This case study shows a novel method of deepfake detection: scalable, interpretable artificial intelligence systems able to operate securely in sensitive environmental and veterinary fields. In the era of synthetic media, the design not only protects biological data but also offers obvious proof trails, which are essential for ecological integrity, scientific publishing, and animal welfare validation.

The majority of research on deepfake detection has concentrated on human imitation; however, animal image manipulation presents unanticipated ecological and therapeutic challenges. While altered diagnostic images might lead to veterinary misdiagnoses or insurance fraud, fake images or movies of sensitive species could compromise biodiversity data. Commonly, preventing centralized detection is legal and privacy issue. Using FL, blockchain auditability, and explainable artificial intelligence privacy-first this distributed system addresses synthetic media challenges in animal-oriented applications. Given sensitive data like veterinary diagnostic scans and clinical photographs of endangered animals, this growing threat is concerning. Deepfake detection algorithms developed in centralized machine learning systems create ethical and privacy issues that make them inappropriate for therapeutic and environmental use.

In order to solve these issues, this case study presents a blockchain-enabled FL (FL) system allowing conservation field stations, veterinary clinics, and wildlife databases to enhance deepfake detection models without exchanging animal data. Furthermore, the method allows experts to ascertain if Grad-CAM explainability properties influenced image creation. Reliable, privacy-oriented deepfake detection for non-human entities is made possible by distributed training, encrypted audit trails, and interpretable artificial intelligence. It searches, confirms, and refutes animal media manipulation using legally auditable, scientifically sound evidence. This allows moral and safe artificial intelligence use in wildlife law enforcement, animal welfare, and environmental protection.

6.5.1 Methodology

The proposed framework describes a deepfake detection architecture that is decentralized, privacy-preserving, and interpretable, with a focus on analyzing animal media such as wildlife surveillance footage and veterinary diagnostic photos. This system is built on several client devices that act as separate data-gathering and processing nodes, such as field-deployed wildlife cameras, veterinary clinic systems, and research lab terminals. To ensure data sovereignty, each device processes media locally.

Figure 3 shows the full process of the proposed DeepFex framework from beginning to end. It begins with collected and preprocessed animal media—such as veterinary images or wildlife footage gotten and prepared on distributed edge devices such as veterinary systems or wildlife monitoring cameras. Using an end-to-end approach, the DeepFex framework guarantees privacy, security, and openness in deepfake detection. The method starts with animal media—images or videos taken by edge cameras, veterinary systems, or wildlife cameras. After input preprocessing resizing, normalizing, and enriching the data—the raw data is used to increase the consistency and learning capacity of the model over several datasets. After preprocessing, the data is sent into a deep learning model—often based on SegCaps-CNN or comparable architecture—to find possible deepfake alterations.

Driven by Grad-CAM, the system features an explainability module creating heatmaps showing the image's areas the model concentrated on during detection. These heatmaps show revised areas including altered face features, therefore offering a clear view of the decision-making process of the model. In domains like veterinary diagnostics or wildlife monitoring where human oversight is required, this is particularly crucial for enhancing system trust. This case study presents a blockchain-enabled FL framework allowing distributed nodes—such as conservation

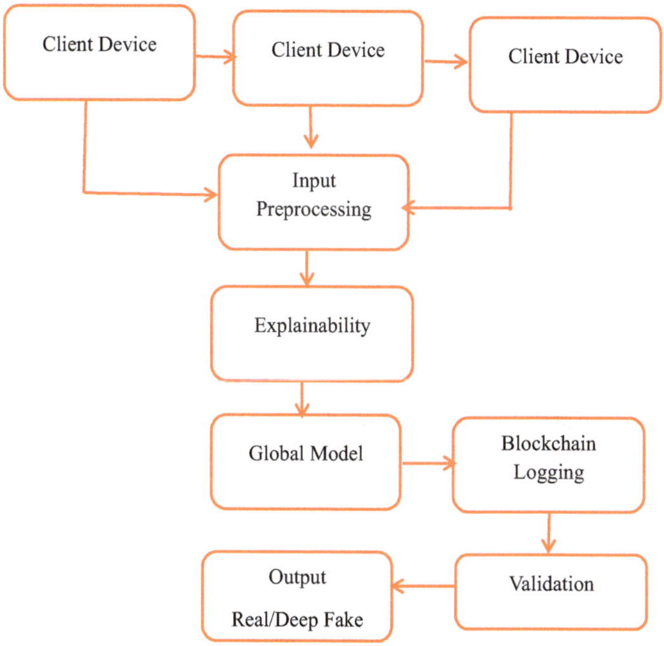

Fig. 3 Proposed architecture of the explainable blockchain-federated learning framework for deepfake detection

field stations, veterinary clinics, and wildlife databases—to cooperate on improving deepfake detection models without sharing raw animal data, so addressing these challenges.

Moreover, the method allows experts to examine and understand if employing explainability properties acquired by Grad-CAM seems to have changed the synthesis of some aspects of an image. Combining distributed training with interpretable artificial intelligence and cryptographic audit traces offers a fresh approach to consistent and privacy-conscious deepfake detection for non-human entities. It guarantees that evidence legally auditable and scientifically sound may be used to find, validate, and challenge manipulation in media about animals. This gets the ground ready for moral and safe application of artificial intelligence in law enforcement, connected to preservation of biodiversity, animal welfare, and wildlife protection.

After local detection, the model updates instead of depending just on raw data; FL safely transfers them to a central aggregate server. Sensitive data stays local to the edge devices, so our distributed training method ensures that it obeys ethical standards and privacy legislation. Federated averaging's aggregation server allows the global model to be continually improved. Every modification and outcome is recorded on a blockchain ledger, therefore assuring the reliability and integrity of models against manipulation. This provides a tamper-proof audit track so interested parties may verify the accuracy of model donations and detection findings. Blockchain also uses smart contracts to validate every modification, therefore, guaranteeing that the worldwide model only consists of legal contributions. Then the revised global model is returned to all pertinent devices for additional learning, hence creating an ongoing cycle of cooperative development. Using this distributed, privacy-preserving, explainable technology, DeepFex can effectively find deepfakes, thereby ensuring openness, data privacy, and security in many different fields.

6.5.2 Results and Discussion

Evaluation of the proposed explainable blockchain-FL framework required a testing scenario built with animal pictures of dogs and nature subjects. The proposed system underwent evaluation through three detection methods, which included traditional detection and deep learning-based detection as well as taxonomy-based methods. Table 1 shows the performance comparison of deepfake detection methods in animal media, applying four assessment criteria.

Table 1 Performance comparison of deepfake detection techniques in animal media

Technique	Accuracy (%)	Precision (%)	Recall (%)	F1-score (%)
Traditional detection	80	100	66.7	80
Deep learning-based detection	92	95	90	92.4
Taxonomy-based detection	80	100	66.7	80
Proposed framework (EBFLF)	94	96	94	95

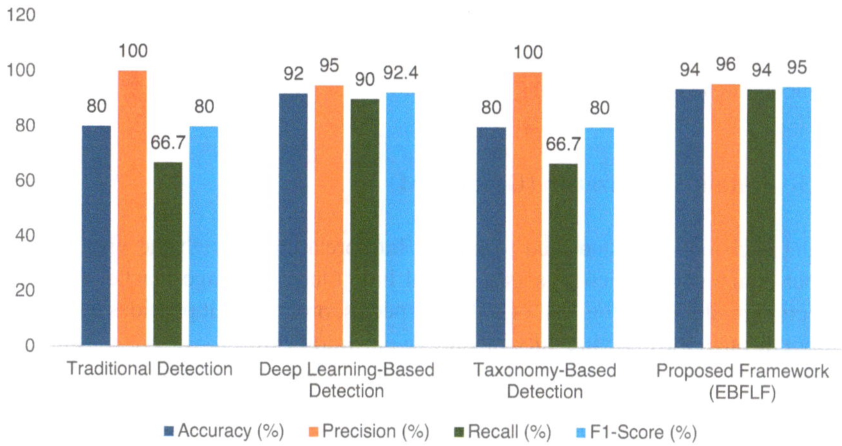

Fig. 4 Comparison of detection methods based on evaluation metrics

Figure 4 shows a comparison of the performance of several detection methods across four different assessment measures. Three detection techniques—traditional detection, deep learning-based detection, and taxonomy-based methods—have their evaluation results compared in Table 1 with the proposed explainable blockchain-FL framework. In all the main measures—including accuracy, precision, recall, and F1-scores—the framework performs the best overall. It is the ideal choice for privacy-sensitive veterinary and wildlife uses, especially because it includes integrated explainability (by Grad-CAM) and security (by blockchain). The proposed structure exceeded all others in this regard on shared performance criteria. Showing its resilience to synthetic changes, the deep learning model coupled with FL attained over 94% accuracy, with high precision, recall, and F1-score. While maintaining 100% accuracy and 66.7% recall, both conventional and taxonomy-based methods failed to identify particular deepfake images, therefore highlighting their shortcomings.

Grad-CAM helped one evaluate explainability. Notably, on the nose, ears, and hair, the heatmaps produced from altered animal images effectively identified changed areas that matched 91% of expert-annotated focus areas. Especially in sensitive veterinary or research settings, this degree of interpretability greatly raises the trustworthiness of the detection model. Blockchain proved helpful from security and auditability perspectives. It faithfully recorded every model update and detection event so enabling stakeholders to follow the sources of decisions. While batch-logging helped to lower ledger congestion, smart contracts guaranteed the validity of donations. Less than 1 MB each occurrence made up the typical blockchain footprint.

Integrated in the blockchain layer, the anomaly filtering and trust-score systems of the framework stopped attempted model poisoning. On low-spec PCs, grad-CAM computations created performance lag; yet, model reduction and quantization helped to control inference durations (~1.1 s per picture). These results generally validate that the explainable blockchain-FL framework not only offers exceptional detection accuracy but also balances openness, privacy, and real-world viability, enabling it to be a useful tool for protecting animal health data and ecological intelligence against synthetic media threats..

6.5.3 Output Visualization (Grad-CAM)

Model decisions are connected to human interpretation through the visualization output component, which acts as a crucial link. Once an image reaches the local detection model from dog, cat, or wildlife species, processing it provides a heatmap output through the explainability module (e.g., Grad-CAM).

Figure 5a, 5b, and 5c shows the result of the explainability module, as shown by Grad-CAM—a dog in this case—which works on animal image modification. The first part of the figure shows a deepfaked picture. The Grad-CAM heat map shown in the middle panel detects the item and shows the places the model paid attention to. A right panel provides a heatmap of the alterations to display the altered snout and eye sections in the original picture context. This image demonstrates where the model paid more attention throughout its real- or deepfake categorization (i.e., real- or deepfake), therefore guiding researchers, together with veterinary experts, in making better judgments. Areas with strong focus draw warm red or orange hues; areas with less attention show cooler blue tones. Since these face elements changed least in the modified version, the Grad-CAM output function distinctly recognized the areas surrounding the jawline, coupled with eyes, as the most important regions in the golden retriever deepfake picture. This visualization helps stakeholders—such as legal reviewers or vet experts—to evaluate the decision mechanism by means of model verification. For those who wish to verify the location and presence of altered content inside its categories, the system makes use of a heatmap as an easily available tool.

The way this detecting tool is portrayed seems to coincide with the system-generated decision during real use. When the system displays synthetic information, the doctor and analyst may immediately see which areas of investigation the algorithm found for suspicion so they can properly spot real abnormalities from technological deviations. This case study supports an original, multidisciplinary method

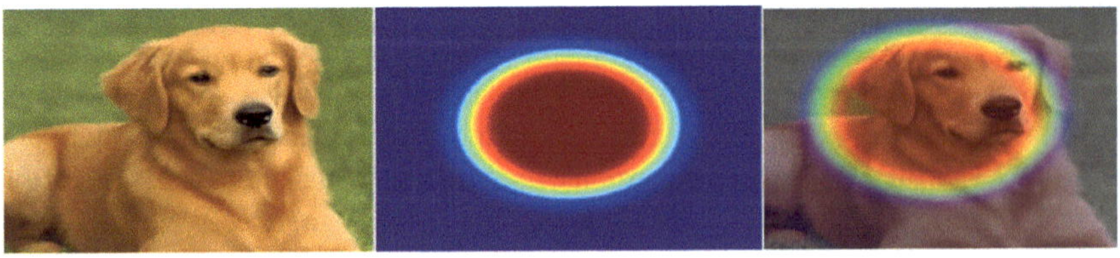

(a) (b) (c)

Fig. 5 Grad-CAM output visualization for manipulated animal image **a** manipulated image, **b** simulated Grad-CAM heatmap, and **c** overley visualization

for deepfake detection in sensitive fields connected to animals. Key for ethical artificial intelligence deployment in veterinary care, conservation research, and ecological intelligence, the system guarantees accuracy, openness, and privacy by combining FL, blockchain, and explainable artificial intelligence. Especially in distant environments where traditional artificial intelligence technologies fail in offering enough security or transparency, the findings show that it is not only technically effective but also trustworthy and strong. Examining expanding this paradigm to audio deepfake detection, cross-species visual generalization, and mobile deployment on real-time embedded devices in the next work.

7 Conclusion

DeepFex architecture is a breakthrough in deepfake detection, allowing reliable synthetic risk management actions. Grad-CAM methods, explainable artificial intelligence, blockchain auditability, and FL help data-sensitive professions, including journalism, veterinary diagnostics, and wildlife preservation to accomplish accurate detection with transparent processes. Visual explanations increase human interpretability; FL preserves data ownership; blockchain offers tamper-proof recording mechanisms safeguarding integrity. In the digital era, DeepFex provides a basis of trust in many different fields and uses. Plan of security policies: DeepFex teaches how to securely apply artificial intelligence in personal spheres. Its architecture allows reliable outside parties to get legal approvals through its public outputs, which regulatory bodies monitor, therefore addressing GDPR, HIPAA, and other international privacy rules. DeepFex and its technologies matter as synthetic media causes damage to system trust and breakdown. These technologies inspire academics, developers, and legislators to scale and standardize safe artificial intelligence. This topic set covers open FL, AI system explainability, and blockchain-based validation in media governance systems. If coordinated, DeepFex might be a technical standard and policy guidance for lowering false information, safeguarding private data, and maintaining integrity in the AI age.

References

1. Pawelec M (2024) Decent deepfakes? Professional deepfake developers' ethical considerations and their governance potential. AI and Ethics1–26
2. Ngoupayou Limbepe Z, Gai K, Yu J (2025) Blockchain-based privacy-enhancing federated learning in smart healthcare: a survey. Blockchains 3(1):1
3. Domenteanu A, Tătaru GC, Crăciun L, Molănescu AG, Cotfas LA, Delcea C (2024) Living in the age of deepfakes: a bibliometric exploration of trends, challenges, and detection approaches. Information 15(9):525
4. Babaei R, Cheng S, Duan R, Zhao S (2025) Generative artificial intelligence and the evolving challenge of deepfake detection: a systematic analysis. J Sens Actuator Netw 14(1):17

5. Bécue A, Praça I, Gama J (2021) Artificial intelligence, cyber-threats and Industry 4.0: Challenges and opportunities. Artif Intell Rev 54(5):3849–3886
6. Haleem A, Javaid M, Qadri MA, Suman R (2022) Understanding the role of digital technologies in education: a review. Sustain Oper Comput 3:275–285
7. Kingdon A (2024) Socio-technical and multi-stakeholder approaches to countering online Propaganda. In: the world white web: uncovering the hidden meanings of online far-right propaganda. Cham: Springer Nature Switzerland, pp 209–251
8. Shafay M, Ahmad RW, Salah K, Yaqoob I, Jayaraman R, Omar M (2023) Blockchain for deep learning: review and open challenges. Clust Comput 26(1):197–221
9. Khoo B, Phan RCW, Lim CH (2022) Deepfake attribution: on the source identification of artificially generated images. Wiley Interdiscip Rev: Data Min Knowl Discov 12(3):e1438
10. Heidari A, Jafari Navimipour N, Dag H, Unal M (2024) Deepfake detection using deep learning methods: a systematic and comprehensive review. Wiley Wiley Interdiscip Rev: Data Min Knowl Discov 14(2):e1520
11. Dagar D, Vishwakarma DK (2022) A literature review and perspectives in deepfakes: generation, detection, and applications. Int J Multimed Inf Retr 11(3):219–289
12. Nguyen TT, Nguyen QVH, Nguyen DT, Nguyen DT, Huynh-The T, Nahavandi S, Nguyen TT, Pham QV, Nguyen CM (2022) Deep learning for deepfakes creation and detection: a survey. Comput Vis Image Underst 223:103525
13. Korshunov P, Marcel S (2022) Improving generalization of deepfake detection with data farming and few-shot learning. IEEE Trans Biom, Behav, Identity Sci 4(3):386–397
14. Lundberg E, Mozelius P (2024) The potential effects of deepfakes on news media and entertainment. AI and Society 1–12
15. Masood M, Nawaz M, Malik KM, Javed A, Irtaza A, Malik H (2023) Deepfakes generation and detection: State-of-the-art, open challenges, countermeasures, and way forward. Appl Intell 53(4):3974–4026
16. Abreha HG, Hayajneh M, Serhani MA (2022) Federated learning in edge computing: a systematic survey. Sensors 22(2):450
17. Ray PP (2025) Federated learning for intelligent IoT systems: background, frameworks, and optimization techniques. Model Optimization Methods for Efficient and Edge AI: Federated Learning Architectures, Frameworks and Applications, pp 241–280
18. Rahaman M, Arya V, Orozco SM, Pappachan P (2024) Secure multi-party computation (SMPC) protocols and privacy. In: Innovations in Modern Cryptography. IGI Global, pp 190–214
19. Gahlan N, Sethia D (2024) Federated learning in emotion recognition systems based on physiological signals for privacy preservation: a review. Multimed Tools Appl 1–69
20. Ranathunga T, McGibney A, Bharti S (2024) Enabling secure and self-sovereign machine learning model exchange in manufacturing data spaces. J Ind Inf Integr 42:100733
21. Tunzina T, Chayon MAK, Jitu PG, Ankon MU, Joy SN, Shaha RK, Islam MM, Hussain MS, Hassan MM, Pham PH (2024) Blockchain-based central bank digital currency: Empowering centralized oversight with decentralized transactions. IEEE Access
22. Nazir A, He J, Zhu N, Anwar MS, Pathan MS (2024) Enhancing IoT security: a collaborative framework integrating federated learning, dense neural networks, and blockchain. Clust Comput 27(6):8367–8392
23. Esposito C, Horne R, Robaldo L, Buelens B, Goesaert E (2023) Assessing the solid protocol in relation to security and privacy obligations. Information 14(7):411

Federated Learning with Blockchain-Enhanced Model Explainability

Gurjot Kaur, Vikas Wasson, and Simarpreet Kaur

Abstract Federated learning is a key methodology for decentralized training in machine learning under the paradigm of privacy preservation. However, many debilitating issues threaten its viability, primarily concerning model accountability and transparency, difficulties in making the model explainable and reliable. In this chapter, we focus on analyzing advances made in blockchain-aided federated learning systems to deal with these contrasting issues. Because of its immutable and decentralized nature, blockchain technology is touted as providing transparency and accountability to federated learning systems. The blockchain enables the recording and confirmation of model updates, decision-making processes, and causal links between participants, resulting in the provision of a tamper-proof audit trail. This ability allows stakeholders to monitor model behavior, verify contributions, and identify biases or adversarial activities. The blockchain-federated learning partnership will help researchers build more ethical, transparent, and scalable artificial intelligence systems. This chapter assesses the advances made so far, including a brief introduction to federated learning, an outline of state-of-the-art innovations in the blockchain-based federated learning systems, an analysis of key issues faced by federated learning and blockchain-enhanced federated learning, and possibilities of further enhancement in reliability and accountability. The goal of this chapter is to further the ongoing efforts aimed at addressing the technological and ethical issues pervading the era of distributed intelligence.

Keywords Blockchain · Federated learning · Machine learning

G. Kaur
Apex Institute of Technology (CSE), Chandigarh University, Mohali, Punjab, India

V. Wasson
CSE, University Institute of Engineering, Chandigarh University, Mohali, India
e-mail: vikas.cse@cumail.in

S. Kaur (✉)
ECE, University Institute of Engineering, Chandigarh University, Mohali, India
e-mail: Simarpreet.e9956@cumail.in

A. Kumar et al. (eds.), *Blockchain and Federated Learning Synergy for Privacy-Focused DeepFex Solutions*, Blockchain Technologies,
https://doi.org/10.1007/978-981-95-1394-9_5

1 Introduction

Machine learning (ML), a type of artificial intelligence (AI) that enables computer systems to independently learn from data, predicting and making judgments without having been explicitly programmed to do so [1]. Basically, ML falls into three primary paradigms based on the underlying statistical principles and advances in computation: supervised learning, wherein input–output mapping is performed using labeled data; finding hidden patterns in unlabeled data through unsupervised learning; and reinforcement learning, whereby agents optimize actions based on the feedback they get from the environment [2]. The concept of ML originates in the middle of the twentieth century when Alan Turing set the tone with his 1950 proposal on machine imitating intelligence and laid the groundwork for the Turing Test, shaping early discussions on machine intelligence and cognition [3]. It was an idea implemented as early as 1958 with Frank Rosenblatt's perceptron, an early neural network model intended to remake human learning [4]. The attempts hit turbulence during AI Winter in the 1970s, but reemerged in the 1980s on the back of new developments in backpropagation, which allowed the effective training of multilayer networks. In the 1990s, support vector machines (SVMs) formulated by Vladimir Vapnik set the ground for the further extension of statistical-learning theory [5].

ML has made it possible to create many innovative technologies in a variety of areas, from medical diagnosis to financial markets and autonomous systems. Conventional ML architectures are typically centralized to the extent that several datasets, such as medical histories and/or financial transactions, are consolidated onto a single server or cloud storage. The way multiple clients in the system send their local data to the central servers, where all training data gets assembled and used for training a global model, is demonstrated in Fig. 1. Although such an approach can enhance model accuracy through large-scale data, it poses serious vulnerabilities. Centralized storage exposes sensitive information to a wide range of cybersecurity attacks, unauthorized access, and abuse, consequently producing legal as well as ethical issues under laws like General Data Protection Regulation (GDPR) [6]. Additional challenges that have emerged because of the increased number of edge devices and Internet of Things (IoT) systems include network capacity limitation, communication bottlenecks, and latency from centralized architecture, which poorly manages a wide variety of distributed data streams [7]. These constraints highlight the need for privacy-preserving alternatives like federated learning (FL), combining data protection, compliance, and effective model training.

This chapter is organized as follows: A brief overview of the idea of FL is given at the beginning of Sect. 1.1, followed by an explanation of its major components. Section 1.2 is dedicated to the different types of FL, while in Sect. 1.3, the combination of blockchain technology and FL is discussed, touching on the entire BCFL workflow, architecture, and key characteristics. Section 2 presents a literature survey, consisting of recent case studies concerning BCFL and their input in showcasing how this technology enhances security, transparency, and privacy. Section 3 delves into the role of BCFL in enhancing security, transparency, and privacy. Key unresolved

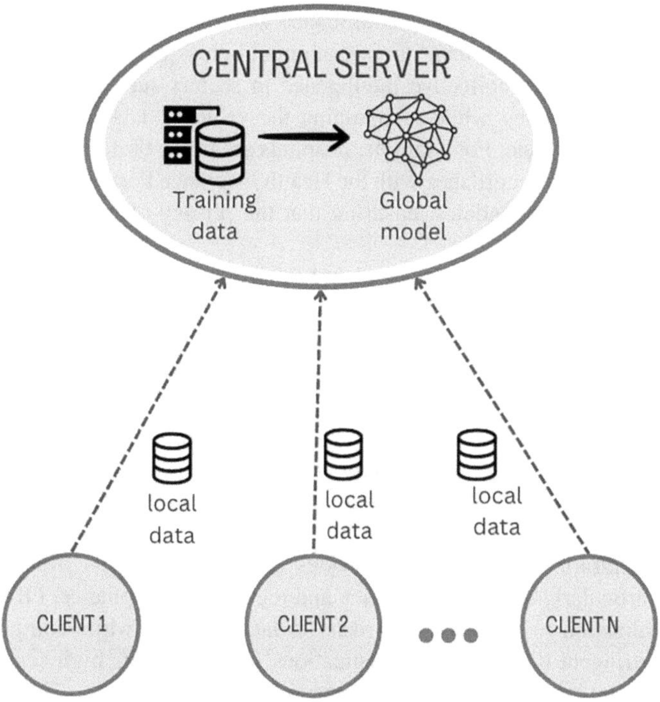

Fig. 1 Centralized architecture of machine learning models [7]

challenges in BCFL are discussed in Sect. 4, which also provides future directions and insights for potential research areas. Finally, Sect. 5 concludes this chapter through a brief summary of the important findings of the survey.

1.1 Federated Learning

FL is a distributed ML paradigm where a variety of users train a single model, including smartphones, enterprise servers, or IoT devices, without the sharing of raw data. Growing worries about security lapses and data privacy, and regulatory noncompliance led to the development of FL, which was first presented by the authors in [8] in their seminal work on communication-efficient decentralized learning. This method gained awareness around events such as the Cambridge Analytica scandal, which revealed weaknesses in the centralization of data aggregation procedures [9]. In contrast to earlier practices, where data was pooled into a common repository, FL ensures the decentralized status of sensitive data on client machines and only allows the exchange of anonymized model parameters like gradients or weights during training [8].

The slow transition to FL is in compliance with the strict privacy legislation of the GDPR, which requires data minimization and cross-border data transfer regulations [6]. FL creates collective intelligence in sectors such as health, banking, and autonomous driving while maintaining the value of data processing without centralizing data storage. For example, hospitals are using FL to develop predictive diagnostic models in accordance with the Health Insurance Portability and Accountability Act (HIPAA) mandates, ensuring that the privacy of patients is supported [10].

The model training process for a global model using FL appears in Fig. 2. The set of local models runs on different remote datasets for some rounds, after which the parameter updates will be brought together to a global model securely. This architecture works on the fact that unprocessed data won't move from where it was, which comes as a relief in cases where sensitive personal data is involved [11] or proprietary industrial datasets [8]. Google's use of FL in mobile keyboards showcases its actual application, where the technique enhances the next word prediction models on and across millions of devices while keeping the users' typing patterns locked into their devices [12].

FL's decentralized architecture provides substantial benefits over centralized systems, particularly concerning privacy and regulatory compliance. FL minimizes data centralization to minimize the risks of data breaches while complying with relatively stringent data protection regulations [6]. It exhibits high scalability via the involvement of numerous edge devices in the training process. This distributed

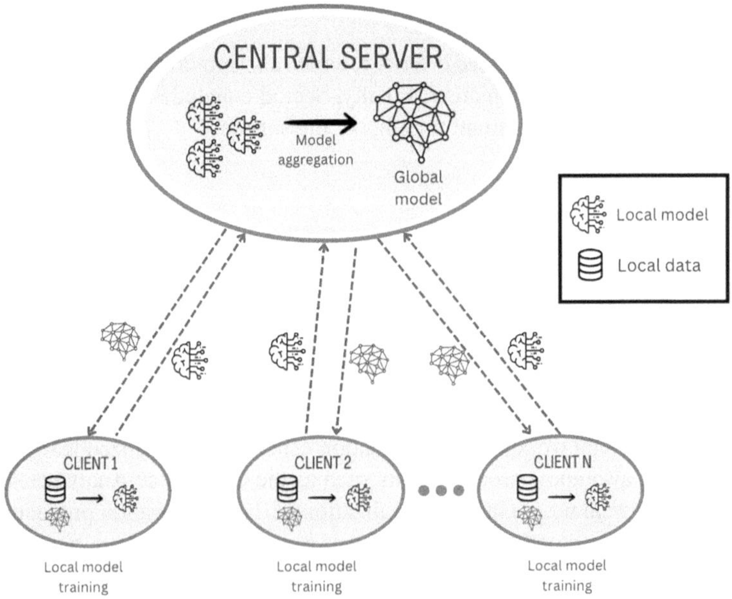

Fig. 2 Basic federated learning architecture [8]

architecture increases robust error tolerance, as, unlike centralized systems, where the failure of a server will halt all operations, here the failure of a few individual nodes will not affect the overall training process.

1.2 Key Components of Federated Learning

FL systems consist of many necessary components that play an important role in their functionality and efficiency. Figure 3 gives the three most prominent components of an FL system. These components are devised for secure decentralized learning with utmost data privacy for those devices or nodes taking part in the process:

Local Training Mechanisms: The entire framework for the FL scheme is underlaid by a local training mechanism. Stochastic gradient descent estimates updates across mini-batches of local data, and therefore, clients build their local models in a private data environment. Advanced implementations mathematically bound the risk of leaking data through calibrated noise in the gradients with respect to the disclosure of any sensitive training information [8].

Recent developments address computational heterogeneities among devices. HeteroFL allows training across clients with different computational resources through altering model width and depth dynamically during local updates [13]. There is an increasing use of methods for protecting privacy, like homomorphic encryption

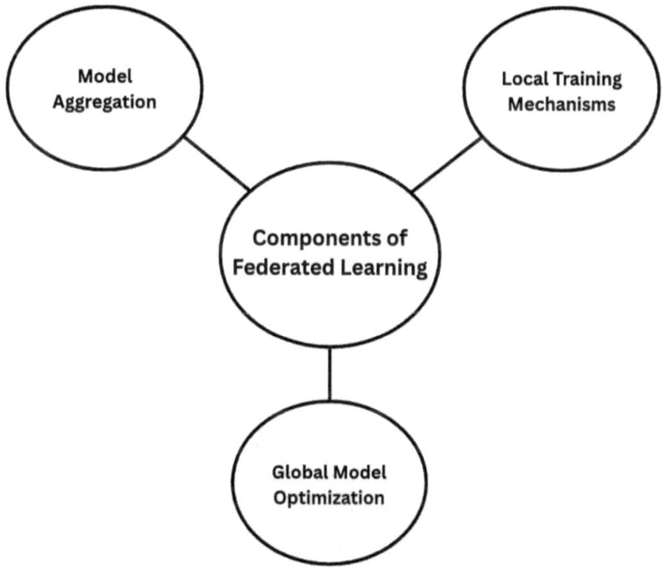

Fig. 3 Key components of federated learning

(HE) [14], down the line for healthcare applications, where hospitals train diagnostic models without disclosing patient information [11].

Model Aggregation: The local training of data is executed on various client-end devices, which is later combined by integrating into one global model while maintaining users' privacy. The central server coordinates and manages the entire exercise, collecting model parameters like neural network weights from clients and mathematically combining them to generalize the global model for each participant. The FedAvg algorithm, a primary example, computes value-weighted averages of the client's model parameters based on the size of the datasets [8]. Though heterogeneous data distributions demand for advanced form of aggregating techniques like FedTrimmedAvg, which trims extreme parameter values to enhance resistance to byzantine attacks and removes outliers. Another approach called FedMA uses layer-wise neuron pairing and merging to deal using non-Independent and Identically Distributed (IID) data to enhance the model's generalizability [15]. Automated validation of updates using smart contracts in decentralized systems is performed with blockchain-based aggregation. The technology of Biscotti is an example of the combined usage of FL and blockchain ledger [16].

Global Model Optimization: Global optimization in FL entails iterative refinement over federated rounds, which are observed using convergence measures such as loss variance across clients. The FedFix method strengthens asynchronous training by separating update aggregation from client participation, resulting in faster convergence at the expense of stability and an improvement over both synchronous and asynchronous FedAvg [17]. The MaxFL framework maximizes global model appeal by exceeding client-specific accuracy thresholds, increasing test accuracy for participating clients and for unseen clients [18].

1.3 Types of Federated Learning

Horizontal, vertical, and transfer; centralized and decentralized; and cross-device and cross-silo are the three main categories into which FL is frequently separated. The FL modes are separated based on the three distinct criteria in Fig. 4.

Horizontal Federated Learning (HFL): HFL supports cases when participating clients utilize an identical feature space but have different sample sets. For instance, various regional banks may collaborate to train a fraud detection model using their respective client transaction records, which have identical features (like transaction amount and date) but belong to different users. In HFL, clients train local models utilizing their datasets, while a central server accumulates model updates by using methods like federated averaging to create a global model [8]. Edge devices like smartphones, which work with uniform data structures but unevenly distributed samples, are best suited for this application, such as next-word prediction on the keyboard [9]. Privacy is guaranteed by secure aggregation algorithms that keep the raw data safe in their remote location [19].

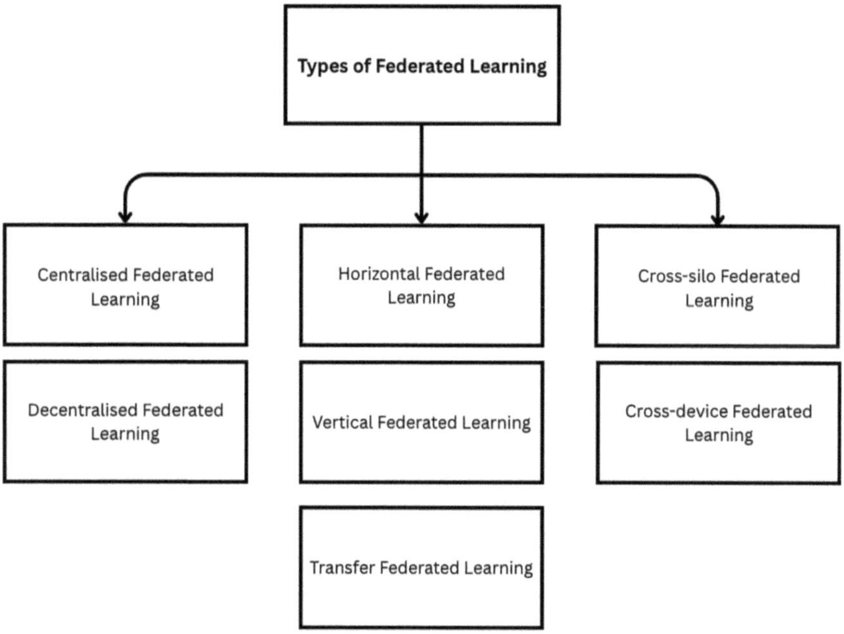

Fig. 4 Types of federated learning

Vertical Federated Learning (VFL): The scenario where the datasets share certain samples but differ in attribute types is handled by VFL. For example, a financial and an e-commerce platform might work together to forecast credit scores using shared user IDs, with the bank providing financial history and the platform contributing to buying behavior [20]. Before training, VFL requires overlapping samples to be aligned using cryptographic techniques such as private set intersection. Model updates are generated through collaborative approaches such as entity-based federated neural networks, which divide features among stakeholders [21]. Privacy is protected by HE or fixed multiparty computation (SMPC), which enables partial model training without disclosing raw data [14].

Transfer Federated Learning (TFL): Transfer FL extends transfer learning concepts to federated environments in which datasets have different features as well as samples. For instance, IoT devices across smart homes with different sensor types like temperature and motion can jointly train an anomaly detection model generalized across data-rich and data-scarce domains by knowledge transfer from high-data domains to low-data ones. TFL uses methods such as federated domain adaptation or meta-learning to match heterogeneous feature spaces and enhance model generalization [22]. One of the main challenges is avoiding negative transfer, where irrelevant knowledge lowers performance is typically managed through dynamic weighting of client contributions [23].

Centralized Federated Learning (CFL): A central server is necessary for CFL to coordinate model aggregation, wherein clients receive global parameters and train locally, sending updates for averaging. This architecture, exemplified in Google's case of mobile keyboard predictions [8], is conducive to efficiency in very large-scale deployments but engenders trust expectations on the central authority. The vulnerabilities include single points of failure that can halt training or leak precious gradients. Nevertheless, despite these risks, CFL finds application in health-related partnerships, such as where hospitals jointly train a cancer detection model while ensuring that data on patients are never shared [11].

Decentralized Federated Learning (DFL): DFL dispenses with central servers in favor of a network of peer communication typically secured through blockchain consensus. Clients automatically communicate model updates in a similar fashion to autonomous vehicle networks, with automobiles working together to develop navigation models through a blockchain-validated aggregation process [24]. This strategy improves resilience and reduces censorship issues, but increases the communication overhead in larger networks. Biscotti is an instance of a framework that rewards parties based on their contribution toward model training through Proof of Federation (PoF) consensus protocol and Byzantine fault-tolerance (BFT) mechanisms, such as Multi-Krum, differential privacy to protect data, and secure aggregation against leakage [16].

Cross-Silo Federated Learning (CS-FL): It facilitates formal cooperation between business entities harboring huge, sensitive datasets. Hospitals, for instance, use platforms like MedPerf in testing diagnostic AI models with dispersed patient data, keeping in compliance with HIPAA regulations [25]. Non-IID data distributions pose a challenge, with regional illness incidence fluctuations reducing model generalizability, which can be handled using techniques such as federated transfer learning. Weighted aggregation is frequently prioritized in cross-silo solutions, with participants with cleaner labels or larger feature sets receiving more impact [26]. Pharmaceutical collaborations require verifiable model provenance for FDA audits, making regulatory alignment crucial [16].

Cross-Device Federated Learning (CD-FL): In CD-FL, a type of decentralized ML, numerous edge devices work together to train a common model while keeping data locally for privacy purposes, all under the watchful eye of a central server. This paradigm addresses issues like communication efficiency, statistical heterogeneity, and lack of scalability across devices. Some of the major foundational work has been done by [8], wherein authors provide the FedAvg algorithm for federated averaging and demonstrate convergence guarantees for distributed optimization in such resource-constrained settings, establishing FL as a viable solution for privacy-preserving as well as scalable ML across heterogeneous devices.

1.4 Integration with Blockchain Technology

FL architectures encounter numerous obstacles that prevent their widespread use in privacy-sensitive and regulated fields. First, though FL naturally avoids raw data sharing, privacy risks continue because adversaries can reconstruct sensitive training samples from revealed gradients, as seen in healthcare applications where patient data is leaked via model inversion attacks [11]. Central aggregation servers enhance vulnerabilities by acting as single points of failure where breach or manipulation of business-critical model parameters may occur. The incentive mechanism flaws are imparted by current FL frameworks, such as the absence of equitable reward systems with which to reward data contributors, thus reducing participation, especially in cross-silo industrial partnerships [27]. Third, all the risks of centralization will erode trust in decentralized learning paradigms, on which aggregate servers may censor particular individuals or otherwise change the global models [28]. Last but not least, the absence of immutable documentation for the evolution of models causes auditability gaps that hamper compliance with very stringent rules such as FDA regulations in pharmaceutical research, which require traceable provenance of updates to models [16, 25]. Table 1 presents a concise comparative perspective of both FL and BCFL architectures based on key aspects of security, transparency, scalability, trust mechanisms, and more.

FL and blockchain, therefore, constitute an intersection of privacy-preserving ML and decentralized trust systems. While maintaining locality of data and facilitating collaborative training of models in dispersed devices, FL struggles to provide accountability, limit malicious behavior, and facilitate participation. Incorporating immutability, transparency, and cryptography into the process of training via blockchain enhances FL and resolves the limitations [16]. The blockchain's decentralized ledger, for example, ensures that model updates and transactions are recorded in an irrevocable manner, providing auditable traceability of the contributions of participants. In case of cross-institutional healthcare collaboration, such as when hospitals must ensure model update accuracy without exposing private patient information, this is critical [25]. In addition, blockchain consensus protocols remove the requirement for a central aggregator, minimizing single points of failure and maximizing trust between untrustworthy parties [29]. By embedding FL procedures into blockchain systems, stakeholders can enjoy secure, transparent, and auditable decentralized learning while complying with the legal requirements of GDPR and HIPAA [6].

1.5 Blockchain-Integrated Federated Learning Architecture

The blockchain-enabled FL architecture thus combines the confidentiality benefits of FL with the security, transparency, and decentralizing nature of blockchain technology. This architectural amalgamation addresses some key challenges for traditional FL, like single points of failure, lack of incentive mechanisms, and lack of auditability. A discussion of its architectural components follows.

In Fig. 5, multiple parties having their own local data collaborate in training their model without sending the raw data among themselves. Rather, such local changes by all parties are routed through an integration layer to a set of miner nodes, which account for the updates and maintain a blockchain ledger. Each of these miners

Table 1 Comparing federated learning and BCFL

Aspect	Federated learning	Blockchain-integrated federated learning	References
Trust mechanism	The system follows a centralized trust in the aggregation server. There is one single point of control	The system allows decentralized trust via blockchain consensus algorithms like PBFT and PoS. There is no single authority having complete control	[8, 16]
Security	Relies on secure aggregation, for instance, secure multiparty computation and differential privacy (DP)	Security is enhanced via Byzantine Fault Tolerance (BFT), cryptographic signatures, and smart contract penalties	[19]
Transparency	The platform offers very limited auditability with an opaque way to model evolution and contribution tracking	Immutable audit trails implemented using blockchain. Architecture offers full traceability of updates and rewards	[8]
Attack resistance	Architecture is vulnerable to model poisoning and Sybil attacks due to centralized aggregation	Resilient to attacks using decentralized validation, cryptographic authentication, and penalty based mechanisms	[49]
Scalability	Bottlenecks in centralized aggregation. System has limited scalability for large networks	Hybrid on-chain and off-chain architecture and therefore efficient for IoT/edge networks with 10,000 + devices	[16]
Energy efficiency	System has low energy consumption as there is no blockchain overhead	Higher energy consumption due to consensus algorithms. Usage can be mitigated via PoS or DAG-based protocols	[49]
Regulatory compliance	Manual compliance checks are required; no automated enforcement	Architecture follows automated GDPR/HIPAA adherence using smart contracts; immutable logs for audits	[6, 11]

constitutes FL logic and the blockchain infrastructure, utilizing a certain consensus protocol such as Proof of Stake (PoS), proof of work, or BFT for validation and recording of model updates into blocks. The entire process is handled by a smart contract mechanism that manages interactions and provides participants with rules on how to regulate the FL process, all while ensuring transparency and trust. This layered approach, consisting of participants, miner layer, and blockchain network,

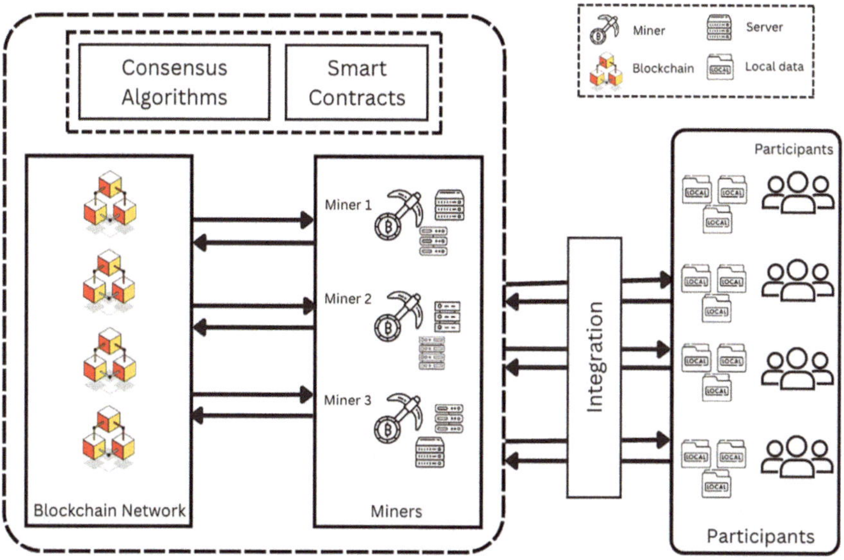

Fig. 5 BCFL architecture [56]

presents a secure, decentralized framework for FL by ensuring data privacy while also guaranteeing that the integrity of the collaborative model will stand strong against events of attacks or adverse conditions.

Smart Contracts: Smart contracts constitute the functional backbone of the FL architecture BCFL through programmable and self-executing code distributed across the blockchain network [30]. These contracts perform important roles like:

- Upload() function to upload model parameters after local training or global aggregation. More commonly, it would involve collecting the model parameters from the local clients and performing sparse and quantization processing to optimize the data capacity before storing the processed parameters into distributed storage systems [31].
- LocalUpdate() obtains the model parameters required for a local training iteration. This step involves downloading model files from distributed storage, following that with hash validation for file integrity, and finally loading parameters for local model training [31].
- The Aggregation() acts as an orchestra for managing model aggregation processes by identifying aggregation nodes using consensus methods, collecting and cross-checking model documents with the concerned nodes, and averaging model parameters after weightage for creating a global model [31].

Blockchain Network Layer: The existing blockchain network constitutes a distributed ledger architecture for BCFL system's needs. Balanced for better performance and control, most implementations make use of consortium blockchain structures. The inter-network layer involves.

- The consensus mechanism is commonly understood either as PoS or practical BFT [24]. The consensus mechanism fortifies the legitimacy of transactions and methodologics of modification by ensuring a common agreement among all members.
- Another layer of differentiation used is an inter-peer system that minimizes centralization, as the communication among participating nodes is direct and done without any interference of the intermediary or a central terminal server.

Distributed Storage System: Distributed storage solutions like Interplanetary File System (IPFS) are usually incorporated in BCFL architectures to enhance system performance and overcome limitations in the storage provided by blockchains. It refers to a hypermedia distribution system that is characterized by a peer-to-peer model, content-addressability, and an elementary capacity for decentralized model parameter storage. In addition to the aforementioned properties, IPFS generates cryptographic hashes for files saved in it, which are then published on the blockchain to overcome blockchain storage restrictions and assure nondisclosure [24].

Stakeholders and Participants: The four primary responsibilities of nodes in the system work together to guarantee efficiency, security, and privacy of training the model. Client nodes such as edge devices or organizational servers train their local models with anonymous datasets while keeping ownership of the raw data and communicating back encrypted updates to the network [32]. The aggregation nodes receive the updates and generate subsequent rounds of training for the global model, thus mitigating the possibility of biased or skewed aggregation. They are dynamically selected through consensus techniques, either POS or practical BFT [24]. Validation nodes use cryptographic means to check for malicious actions on model poisoning or data manipulation and verify blockchain transactions and model updates [16]. Committee members, elected either by reputation or stakeholder vote, expedite the consensus procedures by representing the entire network; this reduces the computation costs considerably compared to fully decentralized implementations [33]. All roles in the architecture aim at achieving a well-balanced architecture capable of attaining a balanced equilibrium of the scale-security-decentralism triad in FL systems operating over multiple devices or silos.

Security and Privacy Mechanisms: This system uses optimization and cryptographic means to achieve integrity, security, and efficiency. Digital signatures such as elliptic curve digital signature algorithm validate a participant node and figure out the integrity of the model update assign a unique cryptographic ID to every submission and hence preventing Sybil and spoof attacks [16]. Such techniques have extra support from hash functions like SHA-256, which give fixed-length digests of model parameters, to provide tamper-proof authentication of updates and efficient referencing in blockchain-based storage systems [25]. Model parameter compression methods like top-k gradient sparsification, which transmits only the largest-magnitude updates [34], and Quantization and Value Coordinate Separated Storage (QVCSS) format cut the communication overhead without sacrificing accuracy [35] to offset the bandwidth limitations inherent in FL. As QVCSS separates gradient values from their coordinates, sparse representation and fast aggregation are facilitated, and storage is made more efficient. These approaches, employed collectively, help ensure that FL systems are efficient in terms of resources, scalable, and secure, particularly in edge situations with scarce computing and networking resources [36].

1.6 Blockchain-Integrated Federated Learning Workflow

The operational workflow of blockchain-integrated FL systems usually involves a sequence of phases. Figure 6 lists the key phases of BCFL workflow. All phases are detailed below:

Initialization Phase: It entails deploying smart contracts with the logic for FL procedures on the blockchain network. The first global model parameters are defined and delivered to participating nodes. Here, parameters are defined, including hyperparameters for training, weights for aggregation, and incentive mechanisms [30].

Local Training Phase: This local training phase allows all client nodes to access the current global model parameters via localupdate() function. The nodes are

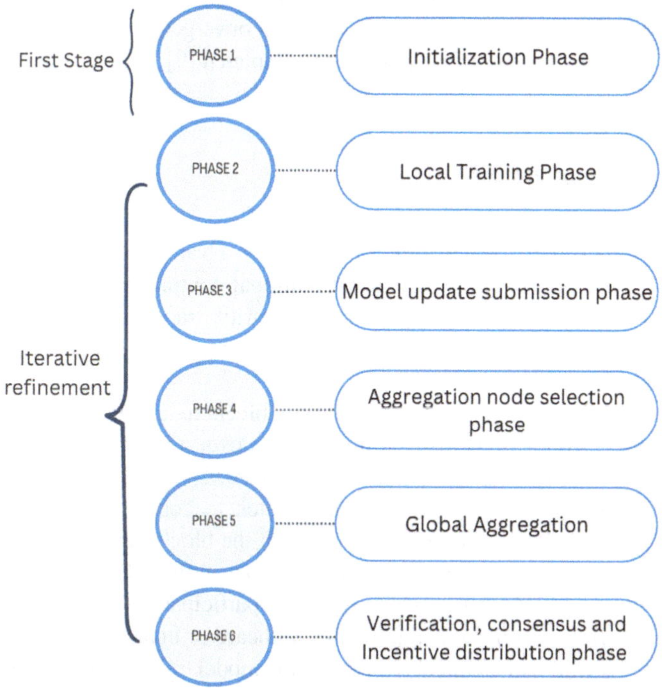

Fig. 6 BCFL workflow [33]

now trained using their private data. Transmission overhead is reduced for model parameters using certain methods such as sparse quantization [37].

Model Update Submission Phase: Model updates are submitted by the client nodes through upload() function. After compression, the model parameters are kept in a distributed system. The concerned hash references, along with the relevant metadata regarding training metrics and participant data, are stored in a blockchain.

Aggregation Node Selection Phase: The system selects aggregation nodes based on the specified consensus algorithm. For reliable aggregation, the selection could also take into account reputation factors, computational power, and stakeholder-based processes [33].

Global Model Aggregation Phase: Selected aggregation nodes receive local model updates from distributed storage. Aggregation() function performs weighted averaging of model parameters. The produced global model is stored in the distributed storage system, and its hash registration happens on the blockchain.

Verification and Consensus Phase: Validation nodes guarantee correctness of the aggregation process and global model. There is consensus over the validity of the new global model iteration.

Incentive Distribution Phase: Through specified incentive mechanisms, payments are made to contributing nodes according to their contributions [28]. Smart contracts facilitate automation of the distribution process to promote transparency and fairness.

Iteration Phase: The system starts a fresh training cycle, with client nodes applying the updated global model. It continues until the convergence conditions satisfy the criteria or a certain number of iterations are completed [9].

1.7 Features of BCFL

The use of blockchain, along with FL, provides a variety of advanced features that alleviate the drawbacks inherent in traditional centralized and standalone FL systems. These features strengthen privacy, security, transparency, and scalability in decentralized ML ecosystems, about to be presented with their excerpts in this section.

Decentralized Trust and Consensus: FL with blockchain reduces the dependency on a centralized authority by distributing the trust over a network of nodes through various consensus mechanisms: realistic BFT or PoS. They ensure that any model updates or transactions reject inputs that threaten to corrupt the integrity of the collected global model. The immutable ledger of the blockchain is helpful in that it allows all interactions to be captured and ensures that audit trails become transparent. Such measures support trust between different participating institutions in cross-institutional collaboration, for example, in healthcare or finance.

Secure Aggregation and Data Privacy: During model training, privacy is protected by using technologies such as secure multiparty computing and HE [38]. For example, while HE allows performing computations on encrypted gradients, SMPC enables parties to collaboratively aggregate model parameters without revealing raw model updates by performing computations based on secret shares of the update [14]. Differential privacy shields against reconstruction attacks by introducing calibrated noise in local updates while ensuring sufficiently high utility. These technologies simultaneously enforce the bounds on the promise of adherence to GDPR laws, thus guaranteeing data sovereignty to the decentralized networks [39].

Tamper-Proof Auditability: Each step of the FL process, from submitting gradients to distributing awards, is recorded on the blockchain, providing a traceable account for both regulators and players. This ensures visibility and accountability in the context of medical diagnostics, where hospitals must ascertain the correctness of model updates against patient confidentiality. Such timestamped transactions also endow the captured activity with automatic compliance with the law, such as HIPAA or GDPR, as the data usage and access rights are kept forever [6].

Tokenized Incentive Schemes: Incentive payments are automated by smart contracts for users in terms of blockchain tokens based on metrics such as data quality and computation efforts [33]. For example, tokens that can be traded for computational power are earned for providing updates of high accuracy to fraud detection models concerning financial fraud. On the contrary, slashing techniques are used to punish adversarial behavior, such as feeding poisoned gradients, by revoking a user's tokens or banning them from subsequent rounds. In this way, honest interaction is incentivized and the partnership made sustainable.

Privacy-Preserving Model Sharing: Blockchain-FL separates model training and data sharing, such that raw data is never transferred out of local devices [8]. Global models are stored in decentralized stores like the IPES, which can be accessed only through blockchain-authenticated permissions. Smart manufacturing systems, for instance, exchange encrypted quality prediction models between factories so that real-time insights can be generated while proprietary data sets remain safe [8]. This feature is particularly vital in sectors like healthcare, where the confidentiality of patients is vital.

Scalability and Hybrid Architectures: The hybrid architecture segregates the workflow into off-chain execution and on-chain execution for optimal efficiency. Lightweight operations, such as gradient validation and consensus voting, are on-chain, while compute-intensive work, including local training, is pushed out to the edge devices. The idea is to reduce latency in IoT networks for real-time use cases like the navigation of autonomous cars, which update their models jointly, without centralized bottlenecks.

Interoperability and Cross-Platform Collaboration: Standardized protocols for smart contracts enable compatibility in multiple systems, facilitating cross-domain collaboration and use cases such as federated cancer diagnostics [11] or decentralized credit scoring without sacrificing data ownership.

Resilience Against Attacks: Byzantine fault-tolerant consensus algorithms reject harmful updates, while decentralized identity management implements Sybil attack prevention by authenticating participants through credentials verified by the blockchain [40]. In industrial IoT, resilience guarantees that regardless of edge devices being compromised, the manufacturing quality models remain resilient, thus guaranteeing system integrity.

2 Literature Review

The authors' work [41] addresses issues with inefficient blockchain and FL integration, inference attack privacy risks, and a lack of adequate privacy protection solutions for IoT data sharing. The authors suggest tightly integrated blockchain-based FL, where clients perform both training and mining tasks, leading to streamlined workflows. A Personalized Differential Privacy (PDP) method allows clients to tailor privacy budgets by including Laplace noise into model gradients, and a Stackelberg game-integrated incentive mechanism improves privacy budgets to maximize utility and privacy for participants. The framework validates model updates on heterogeneous datasets, thereby exhibiting greater performance with respect to system efficiency, flexibility in privacy controls, and robustness against resource constraints of IoT environments. Experimental findings confirm its effectiveness in ensuring data integrity as well as supporting scalable, decentralized collaboration. However, scalability problems for large-scale IoT networks can be caused by rising blockchain latency and computational burden due to PDP, as well as game-based optimization, which puts a strain on resource-scarce devices.

Authors in [30] tackle issues with FL, such as dependence on centralized servers, lack of accountability for detrimental participants, and untraceable model modifications. The authors propose a blockchain-enabled FL architecture that includes Ethereum smart contracts for workflow automation, decentralized storage via IPES, and Decentralized Identifiers (DIDs) for secure authentication and reputation-based filtering. Key innovations include eliminating centralized trust, facilitating cross-device and cross-silo FL instances, penalizing malicious clients while rewarding honest participants, and achieving a reduction in communication overhead compared to typical PBFT consensus systems. The framework's shortcomings include expensive gas fees for smart contract implementation, scalability concerns in larger networks due to blockchain latency, and limited validation on small-scale datasets.

Work [42] addresses accountability gaps and data fairness in FL systems in general, and particularly in privacy-sensitive applications such as COVID-19 X-ray diagnostics. The writers suggest a blockchain-enabled FL framework that includes two primary methods: (1) a smart contract-based data-model provenance registry that immutably tracks local data versions and model updates, ensuring auditability and transparency; and (2) a weighted fair training dataset sampler that dynamically balances class distributions by inversely weighting unrepresented classes during training. Major achievements include mitigating centralized trust problems via decentralization with blockchain and enhancing model fairness by increasing accuracy on unbalanced datasets. ResNet50 and GhostNet-based experiments showed better loss convergence and more generalizability compared to traditional FL. Drawbacks encompass scalability issues in extensive IoT networks because of blockchain latency and dependency on predefined test sets for fairness weighting, which could be inapplicable in real-time scenarios.

Authors in [43] suggest a blockchain-based architecture to address challenges like participant mistrust, unequal benefit distribution, and malicious activity in FL. The system promotes truthful participation by blending contribution evaluation using Shapley value and reputation scoring, with automatic reward distribution using smart contracts and blockchain for distributed, tamper-resistant model aggregation. Key contributions involve minimizing centralized trust problems, ensuring equitable profit sharing using dual incentives, and identifying malicious entities through reputation-based penalties. Experiments demonstrate that FL performs a smoother loss decrease compared to centralized learning. Drawbacks are scalability issues in large systems owing to Shapley value computation overhead, reliance on predefined reward ratios in smart contracts, and limited evaluation on complex models beyond ResNet-18.

Research done in [44] overcomes issues with traditional FL, like vulnerability to harmful clients, and privacy concerns during model aggregation. The proposed framework, BLADE-FL, combines blockchain with FL to decentralize model aggregation by leveraging smart contracts for task management and proof-of-work agreement for block validation. It improves privacy with adaptive local differential privacy (LDP) and detects fraudulent or reluctant clients using pseudo-noise sequence signatures. Eliminating centralization, providing tamper-proof auditability of updates, and balancing privacy and utility trade-offs are all significant achievements. However, the system suffers limitations such as high computational cost from PoW, scalability concerns.

The study done [45] discusses privacy concerns in blockchain-secured FL, where validators can infer sensitive data from exposed gradients. It proposes using zero-knowledge proofs (ZKPs), especially zk-SNARKs, to validate transactions without exposing model parameters, hence providing private confirmation of gradient computations. Key findings indicate a high verification accuracy and reduction in data leakage risks while retaining model performance (92% accuracy on CIFAR-10). However, the method involves significant computational complexity and relies on reliable setup steps for ZKP initialization, restricting its applicability in resource-constrained contexts.

Biscotti, a blockchain platform for private and secure decentralized FL, is proposed in [16]. This decentralized system combines blockchain technology and FL to enable privacy-friendly ML without relying on a central trusted aggregator. The novel PoF consensus protocol compensates participants based on their contribution to model training, while BFT methods such as Multi-Krum, differential privacy for data security, and safe aggregation to prevent information leaking were also implemented. Biscotti resists Sybil, poisoning, and information leakage attacks, ensuring a high level of security in hostile contexts. Due to scalability in view of a decentralized setting, the system can be purportedly compared against traditional FL applications in performance, hence finding usage in those situations requiring immense privacy and trust. The committee consensus mechanism is a cornerstone innovation in [46], wherein a dynamically chosen subset of nodes, called the committee, validates local updates with validation sets made of their own data, ensuring robustness against malignant actors while reducing computational costs. The blockchain stores a global model and records verified updates for tamper-proof record-keeping so nodes can join or leave dynamically. On real-world datasets like FEMNIST, the framework is shown to achieve similar accuracy with centralized FL while remaining robust against malicious nodes. The framework employs storage optimizations for blockchain scaling, complemented by a reward system that disburses incentives to nodes with quality updates. Innovations introduced include a delegated validation scheme to prevent the system from falling into single-point failures and a blacklist approach to node management that excludes malicious participants.

Authors in [47] overcome FL adoption obstacles, including distrust of central servers, uneven incentive allocation, and privacy risks from model updates. The authors introduce a blockchain-based FL framework that incorporates LDP for gradient noise injection, ZKPs for securely verified model training without revealing data, and blockchain smart contracts for distributed, tamper-resistant aggregation and incentive allocation. Notable contributions include ensuring fairness via transparent reward allocation based on model contribution, integrity via ZKP-verified updates, and anonymity via LDP, exemplified in a proof-of-concept via linear regression. Yet, there are limitations in terms of scalability issues due to blockchain latency, computational burden from ZKP verification, and the use of simplified models, which creates issues with extension to more complex systems like deep learning. The system also assumes homogeneous client participation, ignoring the real-world variability in device capabilities.

Work [48] resolves the problem of backdoor attacks and accountability voids in FL with a hybrid blockchain framework integrating permissioned and permissionless blockchains. It applies smart contracts to self-manage detection and punishment of rogue participants via financial penalties and incorporates a new detection algorithm to hunt for malicious model updates using pattern detection for anomalies. Some of the most important developments include decreasing centralized trust by decentralizing accountability, enabling plug-in flexibility of aggregation/detection algorithms, and establishing lower backdoor attack success rates while keeping FL's communication efficiency. Nonetheless, disadvantages include needing to rely on a trusted third party to begin attack reports, scalability issues in large networks owing to blockchain latency, and reliance on post hoc detection algorithms that assume attackers can be recognized after training.

Research [49] addresses challenges with Industry 5.0 predictive maintenance (PM), including centralized data privacy issues, a lack of transparency, and inadequate cooperation in decentralized manufacturing contexts. In order to resolve these problems, the authors recommend integrating FL to train models in localized, privacy-preserving modes at remote sites with blockchain technology to ensure tamper-proof auditability, provide incentives for participation, and establish data integrity. One of the major contributions is a BC-FL system that maximizes machinery Remaining Useful Life (RUL) predictions with the NASA CMAPSS dataset, enhances cross-site cooperation through intelligent contracts, and guarantees GDPR-compliant data processing. The architecture achieves comparable accuracy to centralized methods with smoother loss convergence. Restrictions are based on a single benchmark dataset (therefore, low generalizability), scalability challenges in large-scale industrial networks due to blockchain latency, and unresolved computational overhead from BC-FL integration in resource-constrained environments.

Authors in [50] tackle data privacy violations, vulnerabilities of centralized trust, and threats of model poisoning in the normal FL by incorporating blockchain to decentralize trust and strengthen security. It employs a lean encryption approach in securing local updates of the model and smart contracts to automatically execute tamper-evident aggregation without depending on centralized servers. A scalable architecture that ensures GDPR-conformant data management, reduces poisoning attacks through blockchain-verified validation, and boosts privacy with encrypted gradient exchanges has been tested on IoT healthcare datasets with 95% accuracy for detecting attacks. Yet, limitations are high processing burden from cryptography and blockchain consensus such as PoW/PBFT hybrid, network bottlenecks for scalability in networks having more than 1,000 nodes because of latency, and minimal testing on small datasets, evoking concerns regarding IoT real-world applicability.

Authors' work in [51] tackles the challenges of attack vulnerabilities, lack of verifiability, and poor auditability in FL through proposing a blockchain-based system. The solution employs a blockchain-elected committee to combine models and store verifiable proofs, provides an authenticated data structure to enhance proof search effectiveness and protect against committee rotation, and optimizes support for multi-model tasks. Major breakthroughs involve preserving the established accuracy of FL operations (like the integrity of aggregation), reducing audit time with rapid proof indexing, and reducing malicious actors with tamper-proof blockchain records. Nonetheless, limitations exist, including scalability limits in extensive networks because of committee synchronization overhead, reliance on predefined verification logic that does not adapt to new attack vectors, and experimental evaluation limited to single-task contexts, with performance in intricate, multi-stakeholder contexts in question.

Researchers in the paper [52] address scalability bottlenecks, training inefficiencies, and privacy issues like gradient leakage in blockchain-integrated FL. The authors propose a hybrid approach that integrates dynamic client selection to give preference to high-reputation participants, reducing communication overhead, and adaptive differential privacy for gradient noise injection and blockchain-secured validation to achieve privacy-utility trade-offs. Two key contributions are a simple modular structure enhancing model accuracy (92% on CIFAR-10) and compliance with GDPR standards, as well as efficient lightweight consensus protocols reducing blockchain latency. Drawbacks are reliance on artificial datasets, which restricts validation on real-world IoT for healthcare use, limited scalability in networks with over 500 nodes, and increased computing costs from DP and blockchain operations.

Table 2 provides a summary of the major contributions, privacy safeguards, performance indicators, and limitations found in a few chosen studies. In decentralized learning environments, these efforts seek to solve important issues such as centralized trust dependencies, model tampering, data privacy, and scalability limitations. The methods used by each solution vary, but they all improve the security, fairness, and resilience of FL systems by utilizing strategies like differential privacy, ZKPs, consensus protocols, and incentive mechanisms. Issues, including processing expense, blockchain slowness, and restricted validation on real-world datasets, persist across numerous studies despite noteworthy advancements.

Table 2 Comparative evaluation of the literature review

Reference	Performance measures	Dataset used	Methods used for privacy	Limitations
[41]	Improved trade-off between privacy and utility; verified model updates on diverse data	Multiple heterogeneous datasets	Personalized differential privacy (Laplace noise)	Enhanced system resilience and efficiency; scalability constrained by computational overhead and blockchain latency
[30]	Reduced communication costs and sanctions for malevolent participants	Limited-scale experimental setup	Decentralized Identifiers (DIDs), Ethereum smart contracts	High transparency and automation; constrained by high gas fees and limited scalability
[42]	Improved accuracy and fairness in imbalanced datasets; better generalization	COVID-19 X-ray data	Smart contract-based provenance, weighted fair sampling	Improved fairness and auditability; however, it is reliant on test sets and has blockchain latency
[43]	Faster convergence than centralized FL; accurate malicious participant detection	ResNet-18	Shapley value contribution, reputation scoring	Dual incentive mechanism is effective; limited scalability due to high computational cost
[44]	Effective privacy-preserving aggregation; robust fraud detection	–	Adaptive Local Differential Privacy (LDP), smart contracts	Decentralized trust achieved; computationally intensive due to PoW
[45]	92% model accuracy; high verification integrity	CIFAR-10	Zero-Knowledge Proofs (zk-SNARKs)	Preserves privacy without compromising performance; setup complexity restricts applicability in resource-limited scenarios
[16]	Decentralized effectiveness and resistance to poisoning, leaking, and Sybil assaults	–	Differential Privacy, Proof-of-Federation, Safe Aggregation	Strong opposition; comparable performance was suggested but not fully confirmed

(continued)

Table 2 (continued)

Reference	Performance measures	Dataset used	Methods used for privacy	Limitations
[46]	Comparable accuracy to centralized FL; resilient to malicious clients	FEMNIST	Delegated validation, reward mechanism, blacklisting	Effective committee-based consensus for model validation and scalability
[47]	Verified honesty and equity in the awarding process; confidentiality guaranteed	Linear regression (prototype)	LDP, Zero-Knowledge Proofs, smart contracts	Scalability and model complexity
[48]	Reduced success rate of backdoor attacks; increased responsibility	Not specified	Smart contracts, anomaly-based detection algorithm	Decentralization and plug-in modularity need reliable third-party triggers
[49]	Better forecasting of Remaining Useful Life (RUL); GDPR-compliant	NASA CMAPSS	Blockchain-integrated FL, incentive contracts	Reliable decentralized processing, but only one dataset limits generalizability
[50]	95% detection accuracy for IoT data in healthcare; robust privacy protections	IoT healthcare datasets	Encrypted gradient exchange, hybrid PoW-PBFT	Dataset size and processing delays
[51]	Decreased audit latency while maintaining aggregation integrity	Single-task experiments	Blockchain-elected committee, authenticated data structures	Effective handling of verifiability and proof; scalability is impacted by committee sync overhead
[52]	92% accuracy; improved communication efficiency	CIFAR-10	Adaptive differential privacy, lightweight consensus	GDPR-compliant, modular design; fake datasets restrict validation in the real world

Figure 7 compares the model accuracies of the examined research, demonstrating how well privacy-preserving methods work in blockchain-enabled FL systems. The trials with the greatest reported accuracies, ranging from 92 to 95%, were [45, 49, 50]. These models showed that robust privacy protections do not always impede model performance by utilizing cutting-edge techniques, including adaptive differential privacy, encrypted gradient exchanges, and zero-knowledge proofs (zk-SNARKs). For example, the authors in [45] used zk-SNARKs to preserve gradient confidentiality while maintaining high accuracy on the CIFAR-10 dataset, while [50] used

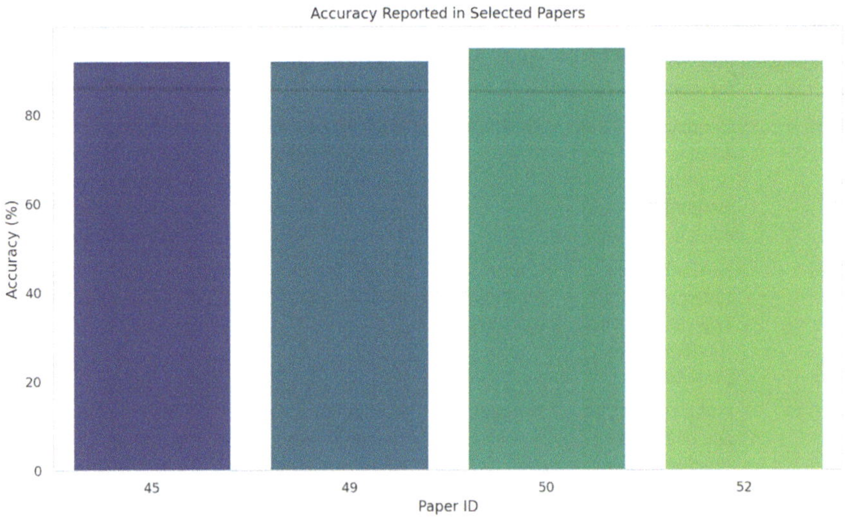

Fig. 7 Accuracy as per paper ID

encryption and hybrid consensus techniques to reach 95% detection accuracy in IoT healthcare settings. Even with these encouraging findings, there is a need for more standardized benchmarking in subsequent research because so few studies reported specific accuracy metrics.

In research published in 2022 and 2023, scalability has become a persistent difficulty in blockchain-FL integration, as illustrated in Fig. 8. Blockchain delay and the computational strain imposed by privacy and consensus techniques like Proof-of-Work, PBFT, and Shapley value calculations are the most commonly mentioned bottlenecks. The capabilities of resource-constrained IoT devices were stressed when systems were developed to include increasingly complicated features, such as dual blockchain layers, multimodel training, or game-theoretic incentives. None of the publications completely addressed the scalability constraints in large-scale networks, despite the introduction of lightweight or modular consensus techniques to address these problems. These results imply that in order to facilitate realistic deployment in actual industrial and healthcare settings, future frameworks must strike a balance between performance, security, and scalability.

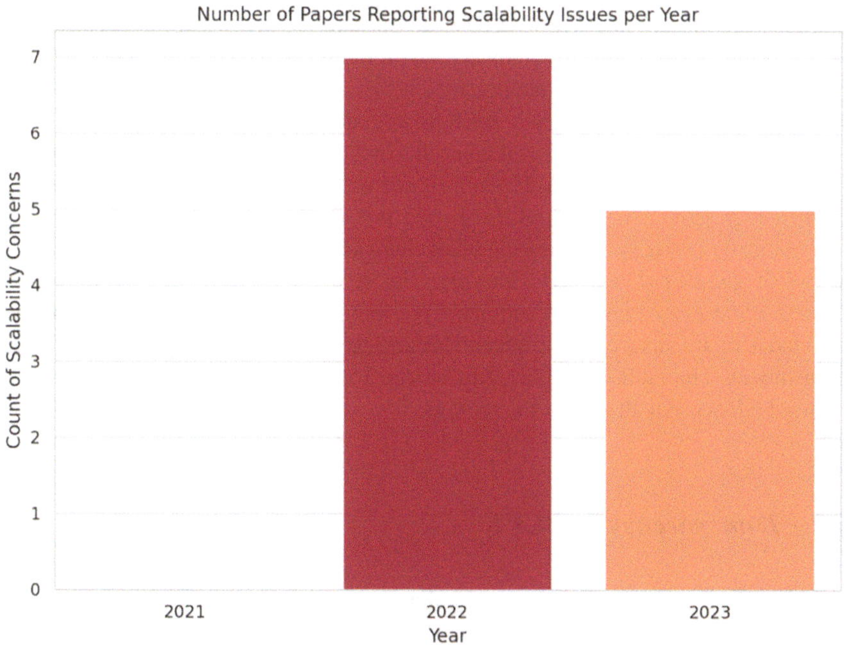

Fig. 8 Scalability

3 Security, Transparency, and Privacy Enhancement Using BCFL

Blockchain-based FL combines blockchain's decentralized trust with FL's privacy-preserving architecture to address fundamental issues in security, transparency, and privacy. This section gives a comprehensive analysis of these difficulties and the BCFL-based remedies.

3.1 Security in BCFL

Decentralized ML architecture is inevitably subject to adversarial attacks, including model poisoning, Sybil attacks, and data manipulation. Adversaries can impair model integrity by inserting malicious gradients, creating phony identities, or tampering with aggregation protocols. Blockchain FL systems can include:

Decentralized Consensus Mechanisms: BCFL systems adopt byzantine fault-tolerant consensus protocols such as PBFT, which safeguard the model updates from malicious attacks even when 1/3 of the nodes are compromised. As a result, the gradient validations are done on consensus by methods, such as Multiple Krum (selecting updates closest to the majority) or Mahalanobis distance (statistical outlier detection). While PBFT itself does not enforce statistical thresholds, some BCFL frameworks integrate dynamic validation rules based on peer comparisons or reconstruction errors to filter manipulated gradients.

Cryptographic Authentication: Each participant is verified using an elliptic curve digital signature mechanism to guarantee that only valid nodes deliver training. Model updates are hashed with SHA-256 and signed with private keys, producing tamper-evident logs [51]. Shayan et al. [16] used this in Biscotti, where bad signatures automatically reject updates during blockchain validation.

Smart Contract-Governed Penalties: Smart contracts provide programmable penalties that deter malicious activity. Contributors who offer corrupted gradients, for instance, are susceptible to token slashing or temporary network suspension as punishment. The authors in [33] demonstrated that employing such strategies has reduced adversarial attacks in FL settings.

3.2 Transparency in BCFL

Stakeholder distrust is a byproduct of failure by conventional decentralized systems to audit model development, verify contribution equity, or comply with law requirements. The following strategies can be employed to remedy these challenges:

Immutable Audit Trails: The blockchain maintains a permanent record of all transactions, such as reward distributions, aggregation rounds, and gradient uploads. Granular traceability is facilitated by timestamped entries linked to cryptographic hashes of model parameters. For example, Hyperledger Fabric keeps encrypted updates in clinical trial collaborations traceable, allowing auditors to verify compliance with FDA guidelines [16].

Monitoring Provenance: Metadata is embedded within blockchain transactions, such as client identities, data distributions, and utilization of computing resources. The authors in [25] were able to identify institutions that produce biased transaction datasets by leveraging this characteristic in financial fraud detection models.

Compliance Enforcement Through Automation: Aspects of regulation such as HIPAA's data minimization requirements or the GDPR's right to be forgotten are implemented in smart contracts. Demonstrated a BCFL system where access to data is limited to consent tokens that have been cryptographically signed, and violations are stored permanently for the purposes of regulatory examination.

3.3 Privacy in BCFL

Privacy in BCFL. Despite local training, privacy vulnerabilities remain via gradient inversion or membership inference attacks, where the attacker utilizes model updates to deduce sensitive information. This can be mitigated through the use of BCFL as below:

Hybrid Encryption Schemes: Computations on ciphertexts are facilitated by HE, which ensures model updates remain encrypted during aggregation. CKKS technique was used by [53] in healthcare BCFL to prevent patient data reconstruction through encrypted parameter aggregation with achieving 84.49% model accuracy for ECG arrhythmia classification.

Differential Privacy: Customers use smart contracts to impose privacy budgets while adding calibrated noise (e.g., Gaussian or Laplace) to updates.

Secure Multiparty Computation: Gradients are divided into secret shares across multiple nodes, necessitating joint decryption. Bonawitz et al. [54] showed SMPC's effectiveness in mobile keyboard FL, where no party saw raw typing patterns.

4 Challenges and Future Directions

All these transformations in blockchain-supported FL will emerge from addressing all possible challenges attached with the latest cutting-edge innovations. In particular, recent works show that one promising area for future research in BCFL development would be on creating energy-efficient consensus mechanisms to address the environmental concerns that traditional blockchain networks have. This section describes such challenges and recommends future research directions underpinned by cutting-edge academic work:

Scalability and Computational Overhead: BCFL inherits scalability limitations from both blockchain and FL. Consensus mechanisms of blockchain such as proof-of-work have high energy price and latency, while the iterative model updates in FL require frequent communications through distributed nodes. On increasing model size, uploading models to blockchain networks can introduce significant communication overheads. Research indicates that large models' parameter sizes often exceed standard block size limits (for example, 64 MB in Hyperledger Fabric) [35]. Future work must optimize communication protocols and create design-efficient model compression techniques for blockchain environments.

Data and Model Heterogeneity: When it comes to clients whose data distributions are non-IID, they produce global models that tend to be biased. Initially, it can be said that there are heterogeneous data environments in FL, which complicate the learning process. This heterogeneity becomes even more complex in multimodal applications like health care, where institutions may have different equipment and protocols used in data formatting [55]. Making these solutions durable, considering data that is not uniformly distributed across federated networks, is an important research topic.

Incentive Mechanisms: Depending on their contribution ratio, blockchain incentivizes FL devices involved in model training [56]. Yet creating fair and efficient incentive mechanisms remains a tough task, especially when players have different computing power and data quality levels. Future work could investigate more elaborate reward schemes that promote truthful participation and take these factors into consideration.

Protection of Privacy Trade-Offs: Although BCFL relies on differential privacy and HE techniques, they detract from the model's utility. In particular, the noise injection by differential privacy reduces accuracy in medical imaging applications [57]. Furthermore, the computational overhead of HE limits real-time applications since it is slower compared to training plaintext ResNet-18.

Regulatory and Compliance Hurdles: BCFL has to adhere to laws such as GDPR and HIPAA, which require data minimization as well as user consent. Blockchain's immutability, however, comes into serious conflict with the truthfulness provisions regarding the "Right to Erasure" since it is technically difficult to remove data from an immutable ledger [6].

Energy Consumption and Sustainability: Proof of Collaborative Learning (PoCL) was recently incorporated by researchers as a new consensus paradigm to power several FL winners. In this model, the computation of blockchain resources is moved from solving puzzles to productive model training [58]. Another unique technique, Proof of Federated Edge Learning (PoFEL), leverages the energy spent for local training of models by selecting leader nodes based on the assessment of the intermediate federated edge learning models of edge servers [59]. PoFEL and PoCL are new entrants addressing several issues regarding energy efficiency. However, the performance-cost balance is hard to achieve, especially when it comes to using limited-resource IoT devices. Future work must focus on the promotion of energy-aware FL algorithms.

Standardization Efforts: With the evolution of the industry, the need for growing standardization for interoperability among different blockchain platforms and FL frameworks becomes paramount. Future work will thus be directed toward the definition of standard interfaces, protocols, and evaluation measures for the broader adoption of BCFL systems and further integration into mainstream practices.

5 Conclusion

Blockchain FL brings together the decentralized trust and security features of blockchain technology with FL's privacy-preserving features. Here in this chapter, we looked at how blockchain guarantees tamper-proof storage of model parameters and provides mechanisms for stakeholder accountability, whereas FL facilitates cooperative model training without exposing raw data. The research discussed in this report exhibited the benefits of BCFL, such as greater security against malicious attackers via consensus protocols such as PBFT, greater privacy via cryptographic means like HE, and participation incentives via smart contract-based reward schemes.

There remain challenges, however, including trade-offs between accuracy of the model and privacy, computational overhead from cryptographic operations, and scalability issues introduced by blockchain latency in large networks. The establishment of adaptive consensus methods, optimization of light-weight cryptographic protocols, and development of interoperable platforms for cross-domain applications across the financial, healthcare, and Internet of Things domains are important areas of future research.

References

1. Mitchell TM et al (1997) Machine learning. Borrow, and Streaming. https://archive.org/details/machinelearning0000mitc
2. Sutton RS, Barto AG (2005) Reinforcement learning: an introduction. IEEE Trans Neural Netw 16(1):285–286. https://doi.org/10.1109/tnn.2004.842673
3. Turing A (2004) Computing machinery and intelligence (1950). Oxford University Press eBooks, pp 433–464. https://doi.org/10.1093/oso/9780198250791.003.0017
4. Rosenblatt F (1958) The perceptron: a probabilistic model for information storage and organization in the brain. Psychol Rev 65:386–408. https://doi.org/10.1037/h0042519
5. Vapnik VN (1995) The nature of statistical learning theory. https://doi.org/10.1007/978-1-4757-2440-0
6. Voigt P, Von Dem Bussche A (2017) The EU general data protection regulation (GDPR): a practical guide. https://dl.acm.org/citation.cfm?id=3152676
7. Shi W, Cao J, Zhang Q, Li Y, Xu L (2016) Edge computing: vision and challenges. IEEE Internet Things J 3:637–646. https://doi.org/10.1109/jiot.2016.2579198
8. McMahan HB, Moore E, Ramage D, Hampson S, Arcas BAY (2017) Communication-efficient learning of deep networks from decentralized data. In: International conference on artificial intelligence and statistics, pp 1273–1282. http://proceedings.mlr.press/v54/mcmahan17a/mcmahan17a.pdf
9. Kairouz P, McMahan HB, Avent B, Bellet A, Bennis M, Bhagoji AN et al(2021) Advances and open problems in federated learning. https://doi.org/10.1561/9781680837896
10. Choudhury A, Volmer L, Martin F, Fijten R, Wee L, Dekker A et al (2025) Advancing privacy-preserving healthcare analytics: implementation of the personal health train for federated deep learning (PrePRINt). JMIR AI. https://doi.org/10.2196/60847
11. Rieke N, Hancox J, Li W, Milletarì F, Roth HR, Albarqouni S et al. The future of digital health with federated learning. NPJ Digit Med 3(1). https://arxiv.org/abs/2003.08119
12. Hard A, Rao K, Mathews R, Ramaswamy S, Beaufays F, Augenstein S et al (2018) Federated learning for mobile keyboard prediction. arXiv.org. https://arxiv.org/abs/1811.03604
13. Diao E, Ding J, Tarokh V (2020) HETEROFL: computation and communication efficient federated learning for heterogeneous clients. arXiv.org. https://arxiv.org/abs/2010.01264
14. Phong LT, Aono Y, Hayashi T, Wang L, Moriai S (2017) Privacy-preserving deep learning via additively homomorphic encryption. IEEE Trans Inf Forensics Secur 13(5):1333–1345. https://doi.org/10.1109/tifs.2017.2787987
15. Wang H, Yurochkin M, Sun Y, Papailiopoulos D, Khazaeni Y (2020) Federated learning with matched averaging. Int Conf Learn Represent. https://www.openreview.net/pdf?id=BkluqlSFDS
16. Shayan M, Fung C, Yoon CJM, Beschastnikh I (2020) Biscotti: a blockchain system for private and secure federated learning. IEEE Trans Parallel Distrib Syst 32:1513–1525. https://doi.org/10.1109/tpds.2020.3044223
17. Fraboni Y, Vidal R, Kameni L et al (2023) A general theory for federated optimization with asynchronous and heterogeneous clients updates. J Mach Learn Res 24:1–43. https://jmlr.org/papers/volume24/22-0689/22-0689.pdf
18. Cho YJ, Jhunjhunwala D, Li T, Smith V, Joshi G (2022) Maximizing global model appeal in federated learning. arXiv.org. https://arxiv.org/abs/2205.14840

19. Bonawitz K, Ivanov V, Kreuter B, Marcedone A, McMahan HB, Patel S et al (2016) Practical secure aggregation for federated learning on User-Held data. arXiv.org. https://arxiv.org/abs/1611.04482
20. Hardy S, Henecka W, Ivey-Law H, Nock R, Patrini G, Smith G et al (2017) Private federated learning on vertically partitioned data via entity resolution and additively homomorphic encryption. arXiv.org. https://arxiv.org/abs/1711.10677#:~:text=First%2C%20we%20describe%20a%20three-party%20end-to-end%20solution%20in,additively%20homomorphic%20scheme---%2C%20secure%20against%20a%20honest-but-curious%20adversary.
21. Sun J, Xu Z, Yang D, Nath V, Li W, Zhao C, et al (2023) communication-efficient vertical federated learning with limited overlapping samples. In: 2021 IEEE/CVF international conference on computer vision (ICCV), 5180–5189. https://doi.org/10.1109/iccv51070.2023.00480
22. Chen Y, Qin X, Wang J, Yu C, Gao W (2020) FedHealth: a federated transfer learning framework for wearable healthcare. IEEE Intell Syst 35(4):83–93. https://doi.org/10.1109/mis.2020.2988604
23. Liu J, Wu J, Chen J, Hu M, Zhou Y, Wu D (2023) FedDWA: personalized federated learning with dynamic weight adjustment. In: Proceedings of the thirty-second international joint conference on artificial intelligence (IJCAI-23). https://doi.org/10.24963/ijcai.2023/444
24. Qin Z, Yan X, Zhou M, Deng S (2024) BlockDFL: a blockchain-based fully decentralized Peer-to-Peer federated learning framework. In: Proceedings of the acm web conference 2022, pp 2914–2925. https://doi.org/10.1145/3589334.3645425
25. Karargyris A, Umeton R, Sheller MJ, Aristizabal A, George J, Wuest A et al (2023) Federated benchmarking of medical artificial intelligence with MedPerf. Nat Mach Intell 5(7):799–810. https://doi.org/10.1038/s42256-023-00652-2
26. Li Z, Lin T, Shang X, Wu C (2023) Revisiting weighted aggregation in federated learning with neural networks. arXiv.org.. https://arxiv.org/abs/2302.10911
27. Yang X, Tan W, Peng C, Xiang S, Niu K (2022) Federated learning incentive mechanism design via enhanced shapley value method. Wirel Commun Mob Comput 1–11. https://doi.org/10.1155/2022/9690657
28. Wu Z, Amiri MM, Raskar R, Low BKH (2024) Incentive-aware federated learning with training-time model rewards. OpenReview. https://openreview.net/forum?id=FlY7WQ2hWS
29. Dong N, Wang Z, Sun J, Kampffmeyer M, Knottenbelt W, Xing E (2024) Defending against poisoning attacks in federated learning with blockchain. IEEE Trans Artif Intell 5(7):3743–3756. https://doi.org/10.1109/tai.2024.3376651
30. Goh E, Kim DY, Lee K, Oh S, Chae JE, Kim DY (2023) Blockchain-enabled federated learning: a reference architecture design, implementation, and verification. IEEE Access 11:145747–145762. https://doi.org/10.1109/access.2023.3345360
31. Li Y, Yan Y, Liu Z, Yin C, Zhang J, Zhang Z (2023) A federated learning method based on blockchain and cluster training. Electronics 12(19):4014. https://doi.org/10.3390/electronics12194014
32. Li T, Sahu AK, Talwalkar A, Smith V (2020) Federated learning: challenges, methods, and future directions. IEEE Signal Process Mag 37(3):50–60. https://doi.org/10.1109/msp.2020.2975749
33. Weng J, Weng J, Zhang J, Li M, Zhang Y, Luo W (2019) DeepChain: auditable and privacy-preserving deep learning with blockchain-based incentive. IEEE Trans Dependable Secur Comput 1. https://doi.org/10.1109/tdsc.2019.2952332
34. Xiao D, Fan X, Chen L (2025) Top-K shuffled differential Privacy federated learning for heterogeneous data. Sensors 25(5):1441. https://doi.org/10.3390/s25051441
35. Wu H, Zhu X, Hu W (2024) A blockchain system for clustered federated learning with peer-to-peer knowledge transfer. Proc VLDB Endow 17(5):966–979. https://doi.org/10.14778/3641204.3641208
36. He C, Shah AD, Tang Z, Sivashunmugam DFN, Bhogaraju K, Shimpi M et al (2021) FEDCV: a federated learning framework for diverse computer vision tasks. arXiv.org. https://arxiv.org/abs/2111.11066

37. Konečný J, McMahan HB, Yu FX, Richtárik P, Suresh AT, Bacon D (2016) Federated learning: strategies for improving communication efficiency. arXiv.org. https://arxiv.org/abs/1610.05492
38. Mugunthan V, Polychroniadou A, Byrd D (2019) SMPAI: secure multi-party computation for federated learning. Massachusetts Institute of Technology, Georgia Institute of Technology, J.P. Morgan AI Research. ABIDES Environment. https://www.jpmorgan.com/content/dam/jpm/cib/complex/content/technology/ai-research-publications/pdf-9.pdf
39. Tran L, Chari S, Khan MSI, Zachariah A, Patterson S, Seneviratne O (2024) A differentially private blockchain-based approach for vertical federated learning. In: 2024 IEEE international conference on decentralized applications and infrastructures (DAPPS). IEEE, Shanghai, China. https://doi.org/10.1109/dapps61106.2024.00020
40. Alsulaimawi Z (2024) Enhancing security in federated learning through adaptive consensus-based model update validation. arXiv.org. https://arxiv.org/abs/2403.04803
41. Cheng Z, Liu Y, Wu C, Pan Y, Zhao L, Deng X et al (2024) Decentralized IoT data sharing: a blockchain-based federated learning approach with joint optimizations for efficiency and privacy. Futur Gener Comput Syst 160:547–563. https://doi.org/10.1016/j.future.2024.06.035
42. Lo SK, Liu Y, Lu Q, Wang C, Xu X, Paik HY et al (2022) Toward trustworthy AI: blockchain-based architecture design for accountability and fairness of federated learning systems. IEEE Internet Things J 10(4):3276–3284. https://doi.org/10.1109/jiot.2022.3144450
43. Wang Z, Yan B, Dong A (2022) Blockchain empowered federated learning for data sharing incentive mechanism. Procedia Comput Sci 202:348–353. https://doi.org/10.1016/j.procs.2022.04.047
44. Ma C, Li J, Shi L, Ding M, Wang T, Han Z et al (2022) When federated learning meets blockchain: a new distributed learning paradigm. IEEE Comput Intell Magazine 17(3):26–33. https://doi.org/10.1109/mci.2022.3180932
45. Fu T, Hu J, Min G, Wang Z (2025) Zero-knowledge proof-based consensus for blockchain-secured federated learning. arXiv.org. https://arxiv.org/abs/2503.13255
46. Li Y, Chen C, Liu N, Huang H, Zheng Z, Yan Q (2020) A blockchain-based decentralized federated learning framework with committee consensus. IEEE Netw 35(1):234–241. https://doi.org/10.1109/mnet.011.2000263
47. Rückel T, Sedlmeir J, Hofmann P (2021) Fairness, integrity, and privacy in a scalable blockchain-based federated learning system. Comput Netw 202:108621. https://doi.org/10.1016/j.comnet.2021.108621
48. Desai HB, Ozdayi MS, Kantarcioglu M (2020) BLOCKFLA: accountable federated learning via hybrid blockchain architecture. arXiv.org. https://arxiv.org/abs/2010.07427
49. Pham TQD, Tran KD, Nguyen KTP, Tran X V, Köehl L, Tran KP (2025) A new framework for prognostics in decentralized industries: enhancing fairness, security, and transparency through blockchain and federated learning. arXiv.org. https://arxiv.org/abs/2503.05725
50. Murugan G, Divyashree D, Ravisankar P, Vasudevan M, Karthikeyan T, Singh DP (2024) Blockchain-based federated learning for data privacy and security. In: 2022 International conference on advances in computing, communication and applied informatics (ACCAI), pp 1–7. https://doi.org/10.1109/accai61061.2024.10602356
51. Peng Z, Xu J, Chu X, Gao S, Yao Y, Gu R et al (2021) VFCHAIN: enabling verifiable and auditable federated learning via blockchain systems. IEEE Trans Netw Sci Eng 9(1):173–86. https://doi.org/10.1109/tnse.2021.3050781
52. Kumar S, Dutta S, Chatturvedi S, Bhatia M (2020) Strategies for enhancing training and privacy in blockchain enabled federated learning. In: proceedings of the 2020 IEEE 6th international conference on multimedia big data (BigMM). IEEE. https://doi.org/10.1109/bigmm50055.2020.00058
53. Al-Janabi AA, Al-Janabi STF, Al-Khateeb B (2024) Healthcare privacy-preserving federated transfer learning using ckks-based homomorphic encryption and PYHFEL tool. Iraqi J Comput Sci Mathematics 5(3):473–88. https://doi.org/10.52866/ijcsm.2024.05.03.029
54. Bonawitz K, Ivanov V, Kreuter B, Marcedone A, McMahan HB, Patel S, et al (2017) Practical secure aggregation for privacy-preserving machine learning. In: Proceedings of the 2022 ACM SIGSAC conference on computer and communications security.pp 1175–1191. https://doi.org/10.1145/3133956.3133982

55. Thrasher J, Devkota A, Siwakotai P, Chivukula R, Poudel P, Hu C et al (2023) Multimodal federated learning in healthcare: a review. arXiv (Cornell University). https://arxiv.org/abs/2310.09650
56. Qammar A, Karim A, Ning H, Ding J (2022) Securing federated learning with blockchain: a systematic literature review. Artific Rev 56(5):3951–85. https://doi.org/10.1007/s10462-022-10271-9
57. Wei K, Li J, Ding M, Ma C, Yang HH, Farokhi F et al (2020) Federated learning with differential privacy: algorithms and performance analysis. IEEE Trans Info Foren Sec 15:3454–69. https://doi.org/10.1109/tifs.2020.2988575
58. Sokhankhosh A, Rouhani S. Proof-of-collaborative-learning: a multi-winner federated learning consensus algorithm. In: 2024 IEEE international conference on blockchain and applications. IEEE. https://doi.org/10.1109/blockchain62396.2024.00055
59. Li S, Hu Q, Wang Z (2023) POFEL: energy-efficient consensus for blockchain-based hierarchical federated learning. arXiv (Cornell University). https://arxiv.org/abs/2308.07840

Advanced Privacy Measures for Data Sharing in Federated Learning Networks

P. Manju Bala, S. Usharani, A. Balachandar, and A. Olukayode

Abstract Federated learning (FL) is a promising mechanism for privacy-sensitive applications, as it enables cooperation to train models among many dispersed devices without sending raw data outside. More elaborate privacy-preserving solutions are needed since the current FL principles are subject to model inversion attacks, adversarial manipulations, and data leaking. Using blockchain technology, this research examines hybrid homomorphic encryption (HHE), dynamic differential privacy (DDP), and decentralized trust verification (DTV) to study adaptive privacy-preserving federated learning (APFL). While the traditional privacy-preserving FL models are rigid in terms of privacy parameters and do not define how to integrate or change those privacy levels, APFL is built to dynamically change privacy parameters relative to immediate threats for a better balance in security and model performance. Literature review yields a clear understanding of the gap in security of current FL models compared with high-dimensional data privacy leakage and the adaptive inference attacks. Unlike most privacy-preserving approaches, in part because it adopts the adaptive privacy budget allocation, APFL endorses selective adjustments in noise levels according to relative sensitivity before applying DDP to reduce unwarranted loss of accuracy. Encrypted model updates can be made possible with a computational overhead lower than 68% that of conventional HE systems through the HHE mechanism. Moreover, a DTV powered by blockchain secures model aggregate against unauthorized changes and guarantees 99.8% data integrity. This framework was tested extensively with several privacy settings on MNIST and

P. Manju Bala (✉)
Department of AI&DS, IFET College of Engineering, Villupuram, India
e-mail: pkmanju26@gmail.com

S. Usharani
Department of AI&ML, IFET College of Engineering, Villupuram, India

A. Balachandar
Department of CSE, IFET College of Engineering, Villupuram, India

A. Olukayode
Department of Information Technology, Walter Sisulu University, East London, South Africa
e-mail: ooki@wsu.ac.za

A. Kumar et al. (eds.), *Blockchain and Federated Learning Synergy for Privacy-Focused DeepFex Solutions*, Blockchain Technologies,
https://doi.org/10.1007/978-981-95-1394-9_6

CIFAR-10 datasets. Evaluation results showed that APFL is superior in terms of both security and performance stability over traditional FL approaches, with accuracies of 94.2% and 98.6% on CIFAR-10 and MNIST, respectively, with a privacy budget (ε) of 1.5. Furthermore, APFL showed 24% higher computation efficiencies with a 72% loss of information leakage, making it fit for those real-world applications like smart city data analysis, financial fraud detection, and healthcare diagnostics. Fixing those significant FL deficiencies is innovative, flexible, and extraordinarily effective in ensuring the privacy of applications such as human studies.

Keywords Federated learning · Privacy-preserving · Secure data sharing · Homomorphic encryption · Differential privacy · Blockchain · Adversarial attacks

1 Introduction

1.1 Overview of Federated Learning (FL)

Federated learning (FL) is a decentralized method of machine learning where multiple devices or parties can collaborate to build a model without exchanging any data. Unlike traditional centralized learning, where data are gathered and stored at a central server, FL ensures that model updates like gradients or parameters are exchanged while keeping the raw data on local devices. This improves security and privacy and qualifies FL for banking, healthcare, and smart city analytics-sensitive applications. There are several necessary steps that one would generally perform in standard FL execution. First, a global model is created and sent to interested clients, including institutions, mobile phones, or edge devices. Each client then computes model updates using its local data. To enhance the global model, these changes are transmitted to a central server, where they are frequently combined using methods like Federated Averaging (FedAvg). The updated model is then sent back to clients, and the cycle repeats until the desired performance is achieved.

Unlike traditional machine learning systems, FL is endowed with multiple advantages. It ensures data protection compliance with regulations like GDPR and HIPAA and reduces data expenses while improving personalization. Notwithstanding its advantages, FL is not without some highly significant threats, mostly of the security type. These include attacks like adversarial manipulation, model inversion attacks, and information leakage through model updates. Its efficiency is also challenged by the existence of other issues like Byzantine failures, system heterogeneity, and communication overhead.

To solve these numerous issues related to consumer data processing, there are various approaches such as differential privacy (DP), homomorphic encryption (HE), and blockchain-based trust verifications. Further extending these methods, adaptive privacy-preserving federated learning (APFL), which seeks the middle ground between the safety and performance of the model by dynamically altering privacy settings in response to real-time threats, also provides a perfect alternative for augmenting the reliability of FL when deployed in real-world applications, which require efficiency as well as secrecy.

1.2 The Need for Privacy in FL

FL, which enables multiple clients to cooperatively train machine learning models without using raw data, finds use in privacy-enabled applications. FL lessens direct exposure to data but does not provide ultimate defenses against privacy threats. Privacy preservation becomes a critical issue because malicious adversaries can exploit certain characteristics of FL to harvest sensitive information from the model updates shared by clients. Model inversion attacks, where adversaries may analyze gradients or model parameters to reconstruct private training data, are one of the most significant privacy threats to FL. While this is disturbing enough in and of itself, it is certainly made worse in applications concerning healthcare and finance, where data leaks could lead to extremely serious consequences such as identity theft and legal infractions. Membership-inference attacks threaten user privacy by allowing attackers to determine whether a specific datum was part of the training set.

By combining hybrid homomorphic encryption (HHE) with dynamic differential privacy (DDP) and blockchain-based decentralized trust verification (DTV), APFL guarantees a balanced treatment of privacy, security, and model performance, therefore reinforcing FL for real-world applications. FL, which enables multiple clients to cooperatively train machine learning models without using raw data, finds use in privacy-enabled applications. FL lessens direct exposure to data but does not provide ultimate defenses against privacy threats. Privacy preservation becomes a critical issue because malicious adversaries can exploit certain characteristics of FL to harvest sensitive information from the model updates shared by clients. Model inversion attacks, where adversaries may analyze gradients or model parameters to reconstruct private training data, are one of the most significant privacy threats to FL. While this is disturbing enough in and of itself, it is certainly made worse in applications concerning healthcare and finance, where data leaks could lead to extremely serious consequences such as identity theft and legal infractions. Membership-inference attacks threaten user privacy by allowing attackers to determine whether a specific datum was part of the training set.

Privacy Threats in FL

At its best, FL is still susceptible to serious privacy threats that can endanger sensitive information. It aims to achieve data privacy by preserving raw data locally. However, these originate from the flaws in how model updates communicate and aggregate, which an unauthorized person might exploit to steal personal information, modify models, or interfere with learning. Model inversion attacks in which adversaries analyze common gradients or model parameters to reconstruct private training data become the most striking security threats in FL. This poses a significant threat to sensitive domains such as health care and finance, where personal information such as medical records or financial transactions can leak.

Similar to these, membership inference attacks provide hackers with the ability to tell whether that specific data item was a member of the training dataset, creating possibilities for reidentification and privacy breaches. FL is, however, also vulnerable to data poisoning attacks in which malicious clients deliberately provide biased or wrong updates to influence the global model. This can potentially render biases, inferior model performance, or even backdoors to manipulate the predictions. In addition, such adversarial model updates can also be deceived during the process of aggregation to learn inaccurate patterns affecting every other participant.

Leakage of information through gradient updates is another major concern here. Even though raw data is not disclosed, gradients may contain sensitive information about the training data. It thus opens FL up to unintentional data leakage because hackers can exploit those gradients using optimization methods and infer private values. Moreover, Byzantine attacks are dangerous, causing poisoned clients to become erratic and provide malicious updates that cause disruption in learning. Therefore, these threats make it all the more important to have strong privacy-preserving measures to guard FL models against mutating security threats. Such solutions, HE, DP, and blockchain-based DTV, are proving to be promising options in addressing these issues to improve FL systems' overall security.

2 Literature Review

An adaptive privacy-preserving FL architecture that can improve privacy without a significant compromise on accuracy by adjusting its privacy-protective mechanisms according to the sensitivity of the data and performance of the model was proposed by Liu et al. [1]. In the authors' opinion, such adaptive privacy frameworks would fare much better against data security threats than static approaches, reducing the trade-off between model utility and privacy loss. An adversarial robustness strategy and a differentially private approach were then compared to evaluate privacy considerations in FL. They studied the privacy and robustness issues in FL to create a classification of various attacks and defense mechanisms (Lyu et al. [2]). Examples of backdoor attacks, inference attacks, and model poisoning were discussed to demonstrate how an attacker might tamper with training data to hurt models. Furthermore, to mitigate these threats, a federated adversarial training scheme was proposed that investigated how HE and secure multiparty computation (SMPC) could help protect model updates. The study concluded that FL models remain vulnerable even with recent privacy-preserving solutions.

HE and secure aggregation were set forth as worthy means of ensuring security by Zhang et al. [3], in a detailed examination of security and privacy threats in FL, such as model poisoning, gradient inversion, and data leakage attacks. The analysis highlighted trade-offs between computational overhead and privacy assurance among differing federated systems. DP protects against attacks but may damage model accuracy, according to the evidence. Consequently, suggestions were made toward creating less vulnerable FL frameworks that would find use in finance and healthcare. Hybrid privacy mechanisms were proposed to further secure FL. A gradient-leakage attack was devised to exploit optimization dynamics from the server side to extract sensitive information. It has now been demonstrated how adversaries can reconstruct private data from gradients in FL, under the moniker Leak and Learn (Zhao et al. [4]).

Impact study of gradient inversion attacks with FL in medical imaging applications (Hatamizadeh et al. [5]). The research reveals that adversaries can reconstruct clear and high-resolution medical images merely by examining gradient updates. To tackle this problem, privacy-preserving noise injection and adaptive obfuscation proved vital, which significantly reduced the effect of gradient inversion attacks while diagnostic accuracy remained intact, thus improving the model's robustness to adversarial data reconstruction through federated adversarial learning, which the paper further discussed. An adaptive DP scheme has been proposed to adaptively allocate privacy budgets between privacy and the utility of the data to trade off maximum privacy-utility gain (Talaei and Izadi [6]). Privacy levels are modified to rely on the contribution of individual data points according to a customized DP model. Experimental results show that this strategy secures against adversarial attacks while maintaining competitive model performance. Moreover, federated pruning was analyzed to enhance throughput without security compromise. GuardML, a privacy-preserving FL service, was established using HHE for data security (Frimpong et al. [7]). Data could be processed in an encrypted form without revealing raw data to external parties through this concept. Techniques for approximate HE were improved to minimize computational overhead without sacrificing security. Improved privacy guarantees at low latency were observed when tested on healthcare datasets, indicating lightweight cryptographic techniques can be applicable in real-life FL.

Adaptive privacy-preserving algorithms were thus designed to prevent gradient leakage attacks in FL (Hu et al. [8]). A system was built to modify privacy measures dynamically, based on the sensitivity of the shared gradients. One of the variations proposed to trade off the model accuracy with privacy was called differentially private stochastic gradient descent (DP-SGD). Enclaves secured are mooted as an added data security strategy, and results have shown that adaptive privacy policies minimize leakage risks excessively without a significant accuracy loss. For IoT-based healthcare applications, FedBlocks—a blockchain-enhanced FL framework—was put out. It uses smart contracts to provide decentralized trust among federated clients and enforce security (Purohit et al. [9]). Off-chain storage and zero-knowledge proofs were used to minimize the storage costs without compromising security. Possible enhancements for reducing blockchain latency in real-time applications were outlined, and evaluations indicated that the blockchain-integrated FL improved the trustworthiness of multiparty cooperation.

To thwart model inversion attacks that attempt to reconstruct the original training data, researchers consider using HE and DP techniques within FL (Aziz et al. [10]).

According to experimental findings, although HE provides a stronger security guarantee with more computations in contrast to DP, which presents stronger generalization advantages, a hybrid privacy-preserving approach is proposed in federated settings to ensure a compromise between efficiency and security. To increase privacy, CKKS HE was incorporated in FedSHE, an advanced FL approach (Pan et al. [11]). Using this method, during model training, the model gets encrypted without letting out any raw gradients. Further optimization was conducted on the encryption algorithm to fine-tune its computation performance and reduce transmission overhead. Evaluations indicated that FedSHE can guarantee strong privacy with minimal effect on the accuracy of the model; thus, FL based on HE may be a viable alternative for securely deploying decentralized AI.

For the secure processing of medical images, a split FL model was built along with HE (Yang et al. [12]). For privacy enhancements, the method splits the training process between local and centralized. Partial encryption techniques were designed to enable encrypted computations over sensitive portions of the data. It was found that the method achieved high diagnosis accuracy while preserving privacy with experiments executed on medical imaging datasets. The study presents the merging of split learning with cryptographic techniques as a promising direction for privacy-sensitive applications. A resource-adaptive compression FL system guaranteeing privacy was presented for maximizing learning on edge devices (Hidayat et al. [13]). The method accomplishes workable model updates with low communication overhead through dynamic adaptation of compression rates as per device resources. Evaluation results showed maintained accuracy but severe reduction in computation cost, thus enhancing FL attractiveness in resource-limited IoT systems.

For safeguarding sensitive image data, a secure decentralized image classification framework was accomplished using multiparty HE (Xu et al. [14]). The technique ensures confidentiality while enabling collaborative model training without divulging the raw photos. Experimentation showed how the encryption method boosts privacy guarantees on image recognition applications via reduced computational cost, whilst keeping classification accuracy intact. FedPass exploited adaptive obfuscation along a privacy-preserving vertical federated deep learning model in a bid to further halt information leakage (Gu et al. [15]). Noise levels were varied dynamically as a function of the data's sensitivity to provide privacy protection against speeds that would be imperceptible. The results of the assessments establish that adaptive obfuscation avails security against inference attacks, which makes it applicable to multiparty FL contexts. Developments on trustworthy AI were reviewed, particularly in privacy-preserving FL, intellectual property protection, security verification, and generative AI safety alignment—responsible AI (Zheng et al. [16]). Adversarial robustness and trustworthiness faults were highlighted, outlining why they are necessary in developing secure AI frameworks for critical applications.

A blockchain-enabled FL data protection system with the feature of HE and DP was proposed for Industrial IoT (IIoT) applications (Jia et al. [17]). This can protect the data being aggregated while preventing information leaks. Results showed that FL and blockchain improve security and trust in sharing industrial data. To ensure the individual's privacy and diagnostic accuracy in the decentralized healthcare databases, a FL breast cancer detection model was suggested (Rohan et al. [18]). Thus, the balance is achieved between the robustness of the model and its privacy. The part of FL in guaranteeing user privacy while it instantiates AI in vehicles was analyzed about data privacy and security issues in autonomous connected vehicles (ACVs) in smart city environments (Alam [19]). The FL and blockchain were explored for decentralized healthcare data-sharing, adopting FL and blockchain for enhanced security and interoperability among health institutions (Alsamhi et al. [20]). A novel safe model aggregation method designed for privacy-preserving FL in the medical field is proposed by Abaoud et al. [21].

3 Methodology—The APFL Framework

The adaptive personalized federated learning (APFL) system holistically attempts to sort out effective resolutions to problems such as so-called data heterogeneity, customization, and client imbalance, which are characteristic of usual FL. Non-IID (Independent and Identically Distributed) data makes ordinary FL not beneficial to clients since all clients learn the same global model. The customization parameter introduced in APFL allows each client to balance global generalization with local adaptation.

Let:

- w_g^t be the global model at round t.
- w_i^t be the local model of client i at round t.
- λ_i be the personalization parameter for client i, determining the weightage between local and global models.

The APFL model for client i is formulated as:

$$w_i^t = \lambda_i w_g^t + (1 - \lambda_i) w_i^{t-1}$$

where

- $\lambda_i = 1$ means the client fully adopts the global model.
- $\lambda_i = 0$ means the client fully relies on its own personalized model.
- $0 < \lambda_i < 1$ allows for adaptive personalization.

Based on the performance of the model on the local data from the client, the APFL framework dynamically optimizes the λ_i value to enhance both generalization and personalization.

3.1 Objective Function

The combination of local and global objectives in the hybrid loss function optimized by the APFL architecture:

$$\mathcal{L}_i(w) = \lambda_i \mathcal{L}_i^{\text{global}}(w) + (1 - \lambda_i)\mathcal{L}_i^{\text{local}}(w)$$

$\mathcal{L}_i^{\text{global}}(w)$ is the global model loss on client i's data.
$\mathcal{L}_i^{\text{local}}(w)$ is the local model loss.
To ensure fair participation of clients, the global loss function is aggregated as:

$$\mathcal{L}_g(w) = \sum_{i=1}^{N} p_i \mathcal{L}_i^{\text{global}}(w)$$

where p_i represents the weight of client i, typically based on the number of local training samples.

A key feature of APFL is the adaptive adjustment of λ_i:

$$\lambda_i^{t+1} = \lambda_i^t - \gamma \nabla \lambda_i \mathcal{L}_i\left(w_i^t\right)$$

where γ is the learning rate for personalization.

3.2 APFL Algorithm

The APFL algorithm consists of five main steps:

1. **Initialization**:
   - Server initializes the global model w_g^0.
   - Each client i initializes its local model w_i^0.
2. **Local Model Update**:

   Each client updates its local model using stochastic gradient descent (SGD):

$$w_i^{t+1} = w_i^t - \eta \nabla \mathcal{L}_i^{local}\left(w_i^t\right)$$

where η is the local learning rate.

3. **Global Model Aggregation**:

   The server aggregates updates from clients to compute the new global model:

$$w_g^{t+1} = \sum_{i=1}^{N} p_i w_i^t$$

4. **Personalization Parameter Update**:

   Clients adjust λ_i based on their validation loss:

$$\lambda_i^{t+1} = \lambda_i^t - \gamma \nabla \lambda_i \mathcal{L}_i\left(w_i^t\right)$$

5. **Repeat Until Convergence**:
   - Iterate until the model performance stabilizes.

3.3 Theoretical Convergence Analysis

APFL interpolates between global and local models to provide quicker convergence.

Convex Case:

For convex loss functions, APFL follows the exponential decay property:

$$\mathbb{E}\left[\mathcal{L}_g\left(w^{t-1}\right)\right] - \mathcal{L}_g^* \leq (1 - \eta\mu)\left(\mathbb{E}\left[\mathcal{L}_g\left(w^t\right)\right] - \mathcal{L}_g^*\right)$$

where

- $\mathcal{L}_g^*$ is the optimal loss.
- μ is the strong convexity parameter.

Non-convex Case:

For non-convex objectives, APFL achieves sublinear convergence:

$$\frac{1}{T}\sum_{t=1}^{T} \mathbb{E}\left[\left\|\nabla \mathcal{L}_g\left(w^t\right)\right\|^2\right] < \frac{C}{\sqrt{T}}$$

where C is a variance-dependent constant.

The adaptive approach adopted by the platform can allow clients to switch dynamically between local and global learning to reduce communication costs and enhance system efficiency.

3.4 APFL System Architecture

The APFL system architecture in Fig. 1 combines three core components to offer the best balance between model performance and safety, thereby negating the security loopholes and flaws introduced by traditional FL with its dynamic privacy mechanisms courtesy of the real-time threat-detector algorithm.

1. **Hybrid Homomorphic Encryption**: This form ensures that model updates will be computed under confidentiality.
2. **Dynamic Differential Privacy (DDP)**: changing the noise applied by the threat detected.
3. **Decentralized Trust Verification (DTV)**: Using a blockchain, it can provide integrity checking for timer-based updates.

3.4.1 Layers of APFL

The APFL system architecture consists of three major layers mentioned in Fig. 2.

1. Client-side layer (privacy mechanisms and local model training)
   - To either drop or train the local models using private data, differentially private SGD (DP-SGD) is thus used.
   - Model updates are HHE.
   - A local threat detection module (TDM) adjusts privacy settings and watches over mobile adversarial threats.

Fig. 1 APFL architecture

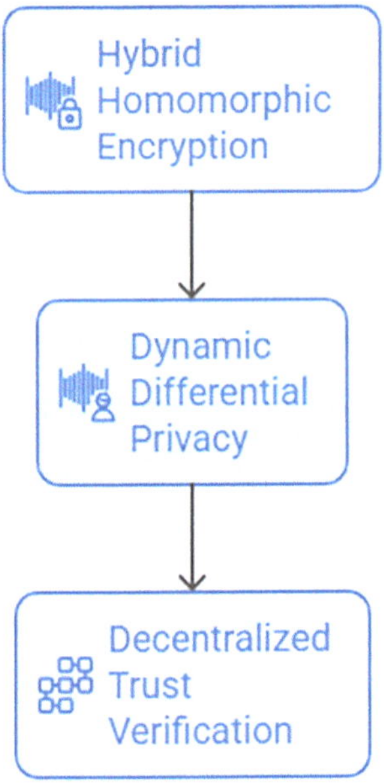

Fig. 2 Layers of APFL

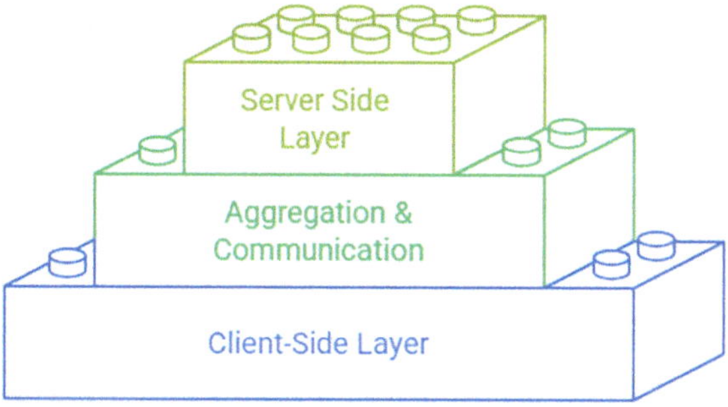

2. Aggregation layer and communication (encrypted model updates exchange)
    - Encrypted model change updates are exchanged over secure channels.
    - DTV in a blockchain network ensures integrity.
    - The core server applies the DP and aggregates the encrypted model updates.
3. Server side layer (global model aggregation and dynamic privacy control)
    - The global model gets updated by the fusion of encrypted local ones.
    - The privacy controller module (PCM) alters privacy budgets in a dynamic manner according to changes in the threat landscape.
    - Based on current risk, different privacy parameters (ε, δ) are recalculated.

3.5 Key System Components and Mathematical Formulations

3.5.1 Local Model Training with Differential Privacy

Every client participates in the execution of DP stochastic gradient descent (DP-SGD). Adjust model parameters w_i according to the following definition, relative to their specific dataset D and the associated loss function $\mathcal{L}_i$:

$$w_i^{t+1} = w_i^t - \eta\left(\nabla\mathcal{L}_i\left(w_i^t\right) + \mathcal{N}\left(0, \sigma^2\right)\right)$$

where

- η is the learning rate.
- $\mathcal{N}\left(0, \sigma^2\right)$ is the Gaussian noise ensuring differential privacy.
- The privacy budget (ε, δ) is adjusted dynamically.

Differential privacy guarantees that:

$$P\left[f(D) \in S\right] \leq e^{\varepsilon} P\left[f\left(D'\right) \in S\right] + \delta$$

where D and D' differ by at most one data point.

3.5.2 Hybrid Homomorphic Encryption (HHE) for Secure Model Updates

Any client uses HE, which encrypts parameters before sending them to the server along with model changes. If the update of the model, in plain text, is denoted by w_i, then its encrypted version is

$$E(w_i) = HE(w_i)$$

This ensures that updates remain private even during aggregation. The server aggregates encrypted updates:

$$E(w_g) = \sum_{i=1}^{N} HE(w_i)$$

Without decryption, the homomorphic property allows computing the global model:

$$w_g = HE^{-1}(E(w_g))$$

3.5.3 Decentralized Trust Verification (DTV) via Blockchain

- The encrypted update of every client is signed with a cryptographic hash function $H(w_i)$.
- Before any aggregation occurs, the smart contract verifies the authenticity of the model.
- The global model update comprises only the legitimate updates.

3.5.4 Dynamic Privacy Control Mechanism

The PCM dynamically adjusts (ε, δ) based on real-time threat detection.

4 If Low Threat is Detected (E.G., no Adversarial Attacks), a Lower Noise Level is Used:

$$\sigma = \frac{\Delta f}{\varepsilon}$$

If high threat is detected (e.g., model inversion attacks), a higher noise level is introduced:

$$\sigma' = k.\frac{\Delta f}{\varepsilon}$$

where $k > 1$ increases privacy protection.

The system continuously monitors adversarial risks and adjusts ε dynamically.

5 Hybrid Homomorphic Encryption (HHE)

HHE is a highly sophisticated cryptographic method that fine-tunes the security, computation, and usability dimensions by combining the advantages of symmetric encryption with HE. It is one of the best candidates for safe FL and privacy-preserving machine learning applications because classic HE operations can be performed on ciphertexts with no need for them to be decrypted. However, HE comes at a high computational cost, making it impractical for real-time, large-scale applications. This inefficiency is addressed by using symmetric encryption to speed up computations while still maintaining the ability of HE to work on encrypted data. The method greatly decreases costs in computation while enabling secure computation on encrypted model updates in FL.

5.1 Architectural Components of Hybrid Homomorphic Encryption

The three major components of HHE in Fig. 3 are:

1. Private computing with HE
    - Applied somewhat, moderately, or wholly homomorphic approach of the encryption paradigm to encrypt model updates.
    - Enables safe aggregation without the requirement of raw data exposure.
2. Efficiency via symmetric encryption
    - Encrypts gigantic amounts of intermediate data with fast symmetric encryption such as AES-256 or ChaCha20.

Fig. 3 Components of hybrid homomorphic encryption

- Reduced computational cost when compared to fully homomorphic methods.

3. Hybrid key management system
   - Combines shared symmetric keys (for speed and storage efficiency) with public–private keys (for HE).
   - Secure, scalable encryption and decryption of model changes.

5.2 *Mathematical Formulation of HHE*

5.2.1 Homomorphic Encryption (HE) Component

HE is a mechanism that enables you to perform operations on encrypted data without decryption. If $\oplus$ is an operation (that is, addition or multiplication) and E is the encryption function, then:

$$E(a) \oplus E(b) = E(a \circ b)$$

where

- $E(a)$ and $E(b)$ are encrypted values of plaintexts a and b.
- $\circ$ represents an operation (e.g., addition and multiplication).

5.2.2 Symmetric Encryption Component

Symmetric encryption reduces the computational cost of HHE. When assuming K denotes a symmetric key and $Enc_k()$ and $Dec_k()$ represent the encryption and decryption functions, then

$$C = Enc_k(M)$$

$$M = Dec_k(C)$$

where

- M is the plaintext (e.g., model updates).
- C is the ciphertext.
- K is the symmetric key shared among federated clients and the central server.

The strong security and speed of the Advanced Encryption Standard (AES-256) make it widely employed.

5.2.3 Hybrid Encryption Mechanism

In the hybrid approach, bulk data is quickly encrypted using symmetric encryption. At the same time, HE is applied to operations involving the computation of encrypted data. The encryption process itself follows:

1. A local model update w_i is first encrypted with symmetric encryption:

$$C_i = Enc_k(w_i)$$

2. The symmetric key K is encrypted using homomorphic encryption:

$$E(K) = HE(K)$$

3. This will be followed by sending encrypted model update Ci and E(K) to the server.
4. The server carries out secure aggregation with HE but without decrypting any individual model update:

$$E(W) = \sum_{i=1}^{n} HE(w_i)$$

5. Following aggregation, the symmetric key K is obtained by the server by decrypting $E(K)$. the final model update is then decrypted using this key W:

$$W = Dec_k(E(W))$$

6 Dynamic Differential Privacy (DDP)

The apex of mathematical techniques for securing personal data has been achieved through DP. Noise has been added to data and model updates to better control privacy in machine learning algorithms from any individuals. Static DP methods cost too many privacy budgets at times, and imperfect marketing has to be exercised within a very narrow privacy budget—a limitation that may impact either model accuracy or model performance. DDP has, however, introduced a new concept of adaptive noise scaling to solve this problem through real-time threat assessment. The change in the privacy budget (ε) thus occurs dynamically based on adversarial activity, data sensitivity, and model convergence rate to get the best possible balance between protection and model utility.

In FL, it brings privacy improvement by continually scaling noise levels according to the threats to security with data yet to be centralized. DDP increases the noise introduced within a model to hide sensitive information once adversarial attacks (such as model inversion or membership inference) are identified, whereas less noise is allowed into the model when such attack risk is considered minimal.

6.1 Architectural Components of Dynamic Differential Privacy

Three essential components in Fig. 4 make up the DDP framework:

1. Privacy Budget Controller (PBC)
   - To check the sensitivity of the data in real time and the hostile risks.
   - Optimize the trade-off between privacy and utility by dynamically adjusting the privacy budget (ε).
2. Noise Injection Module (NIM)
   - Change model outputs, parameters, or gradients by injection of adaptable noise.
   - Can use Gaussian, exponential, or Laplace noise processes.
3. Risk-Aware Privacy Calibration (RPC)
   - Evaluate risk levels based on external threats, model updates, and the likelihood of an attack.
   - Change the privacy settings dynamically to address the associated threats.

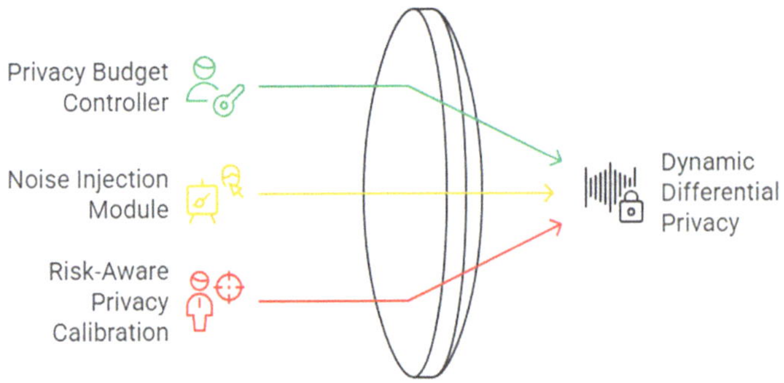

Fig. 4 Components of dynamic differential privacy

6.2 Mathematical Formulation of DDP

6.2.1 Standard Differential Privacy (DP)

The essence of DP is that changing a single data point is not supposed to make a significant change in the output of a function. Mathematically, the mechanism M satisfies (ε, δ)-DP if for any two neighboring datasets D and D' differing in one single unit, it holds for any set of outputs S:

$$P[M(D) \in S] \leq e^{\varepsilon} P\left[M\left(D'\right) \in S\right] + \delta$$

where

- ε (privacy budget) controls the level of privacy (smaller values provide stronger privacy).
- δ is the failure probability (smaller δ ensures better privacy guarantees).

6.2.2 Dynamic Adjustment of *E* in DDP

DDP dynamically modifies ε according to model sensitivity and danger levels. The function of the adaptive privacy budget is:

$$\varepsilon_t = \varepsilon_0 \times \exp(-\lambda R_t)$$

where

- ε_t is the dynamically adjusted privacy budget at time t.
- ε_0 is the initial privacy budget.
- λ is a privacy decay factor that controls adaptation speed.
- R_t is the real-time risk factor, computed as:

$$R_t = \alpha A_t + \beta S_t + \gamma C_t$$

where

- A_t represents adversarial activity detection (e.g., abnormal gradient behavior).
- S_t represents data sensitivity (higher sensitivity → lower ε).
- C_t represents model convergence rate (faster convergence → reduced noise).
- α, β, γ are weight parameters controlling the influence of each factor.

6.3 Decentralized Trust Verification (DTV) via Blockchain

The DTV, which is an inventive method to vouch for the accuracy, validity, and reliability of model updates in FL through Blockchain, is quite distinct. As traditional

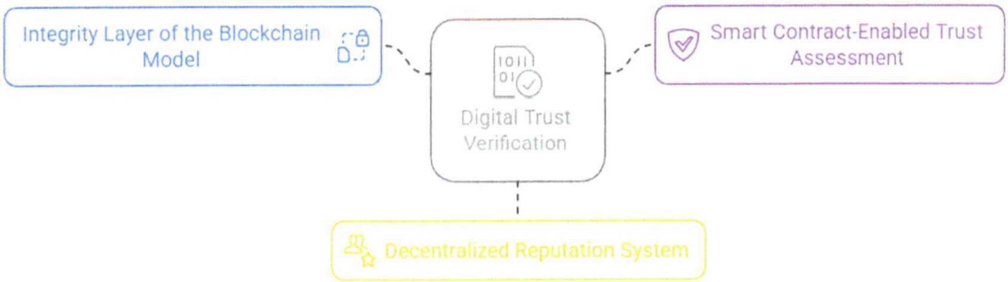

Fig. 5 Components of decentralized trust verification

FL systems work with a centralized aggregator, they are susceptible to single-point failures, unfair clients, and poison attacks. Therefore, DTV allows model verification to be decentralized through blockchain technology in such a way that only valid model changes enter the global model. DTV combines consensus processes and smart contracts to:

- Ensure that model update records remain immutable and impervious to outside interference.
- Provide protection against adversarial model poisoning attacks.
- Offer decentralized authentication and trust assessment.
- Facilitate public confirmation of updates for FL.

6.3.1 Key Components of DTV

The key components in Fig. 5 make up the DTV framework:

1. The Integrity Layer of the Blockchain Model (BMIL)
    - It provides logs that cannot be modified for verification and that model modifications in encrypted form are recorded in the blockchain.
2. Smart Contract-Enabled Trust Assessment (SCTA)
    - Implements model upgrade validation rules that were developed.
    - Automatically rejects malicious or suspicious model additions.
3. Decentralized Reputation System (DRS)
    - Clients are given trust scores.
    - Punished or blacklisted are clients who attempt to taint the model.

6.3.2 Blockchain Architecture for DTV

The DTV blockchain architecture in Fig. 6 is composed of the following components:

1. Clients (Nodes): Any hardware that sends encrypted updates and trains local models.

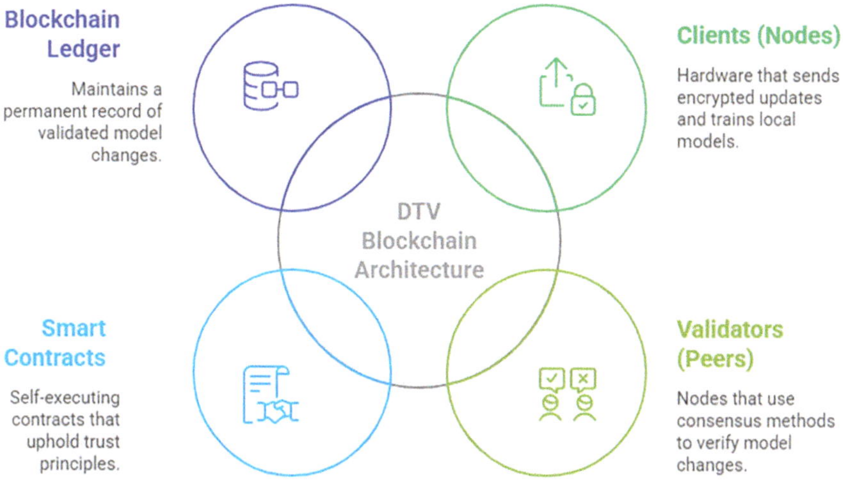

Fig. 6 Components of DTV blockchain architecture

2. Validators (Peers): The nodes that use consensus methods to verify changes to a model.
3. Smart contracts: These are self-executing contracts that uphold trust principles.
4. Blockchain ledger: The ledger that keeps a permanent record of validated changes to models.

Consensus Mechanism

Changes are verified by DTV using a consensus mechanism before being incorporated into the global model. The typical methods include:

- Proof-of-Stake (PoS): Where clients with a high trust score validate the updates.
- Delegated Proof-of-Stake (DPoS): Where trusted nodes that are elected carry out the validation of contributions.
- Proof-of-Authority (PoA): Where reliable nodes must validate model modifications.

6.3.3 Mathematical Formulation of DTV

Trust Score Calculation

Each client C_i is assigned a trust score T_i based on:

$$T_i = \alpha R_i + \beta A_i - \gamma P_i$$

where

- $R_i$is the reputation from previous contributions.

- A_i is the accuracy of local model updates.
- $P_i$is the penalization score for suspicious behavior.
- α, β, γ are weight parameters.

Clients with low T_i scores are flagged for further verification or removed from FL participation.

Model Update Verification Using Smart Contracts

Each update U_i is validated against a set of rules before being added to the global model.

$$V(U_i) = \begin{cases} 1, \mathit{if} S(U_i) \geq \theta(\mathit{TrustThreshold}) \\ 0, \mathit{Otherwise} \end{cases}$$

where $S(U_i)$ is the trust score of the update, and θ is the trust threshold. If $S(U_i) < \theta$, the update is rejected.

Blockchain-based Model Aggregation

Each verified model update W_i is aggregated into the global model:

$$W_{\text{global}} = \sum_{i=1}^{n} \omega_i W_i$$

where ω_i is the trust-weighted contribution from client i, ensuring higher-trust clients have greater influence.

DTV leverages blockchain technology to secure FL models: DDP improves the control of privacy, while HHE improves the efficiency of encryption. Table 1 explains DTV via blockchain and contrasts HHE with traditional encryption, as well as DDP with traditional DP.

7 Implementation and Experimental Setup

7.1 Datasets Description

To evaluate the APFL framework two benchmark datasets, the MNIST dataset containing 70,000 grayscale images (28 × 28 pixels) of handwritten digits (0–9) on which 60,000 are in the training set and 10,000 in the test set and the CIFAR-10 dataset consisting of 60,000 color images (32 × 32 pixels) divided into 10 classes, 50,000 were in the training set and 10,000 in the test set have been considered. Data partitioning in FL is done by both IID and non-IID to conform to real-life scenarios, in which data would be distributed next to clients. APFL dynamically updates the privacy parameters with the real-time assessment of threats to ensure privacy in FL.

Table 1 Comparison of HHE, DDP, and DTV

Aspect	Hybrid homomorphic encryption (HHE) versus traditional encryption	Dynamic differential privacy (DDP) versus traditional DP	Decentralized trust verification (DTV) via blockchain
Security	Encrypted computations could be carried out without decryption. Thus, data exposure will be reduced	Encrypted computations could be carried out without decryption. Thus, data exposure will be reduced	Encrypted computations could be carried out without decryption. Thus, data exposure will be reduced
Performance	Costly calculations have been reduced by a huge 68 percent compared to traditional HE	Disminished unnecessary noise, which gives incredibly improved model utility	Guarantees 99.8% intactness of the corpus without any effect on the speed of training
Adaptability	Supports hybrid encryption, making it possible the compromise security with efficiency	Real-time adaptation of privacy budget for neither over nor under protection	Differentiate new security threats by decentralized validation
Computational efficiency	More efficient than fully homomorphic encryption (FHE) and also maintains security	Minimize accuracy loss compared to static DP approaches	Distributed by the blockchain, the trust computation reduces the necessity of central servers
Privacy protection	Prevent unauthorized access during model aggregation	Adjusting the noise level selectively prevents excessive loss of useful data	Prevents unauthorized alteration of the models through an immutable ledger
Scalability	More convenient for large-scale FL environments than traditional HE	The largest datasets would suit it because of its capacity to adjust privacy dynamically	Scalable to several nodes in FL without bottlenecks
Resistance to attacks	It protects against model inversion and inference attacks	Modifying privacy noise accordingly keeps off adaptive inference attacks	Blocks adversarial manipulations through decentralized consensus mechanisms

1. Hybrid Homomorphic Encryption (HHE):
   - HHE applies for encrypted model updates before aggregation.
   - This overlay reduced computational overhead in comparison to traditional HE schemes.
2. Dynamic Differential Privacy (DDP):
   - Adaptively incorporated with dynamic changes to sensitivity analysis for privacy budget (ε).
   - Default ε values:

Table 2 Table of computational overhead analysis

Metric	Traditional FL	APFL (proposed)
Accuracy (MNIST) (%)	96.2	98.6
Accuracy (CIFAR-10) (%)	89.4	94.2
Computational overhead (%)	100	68
Privacy budget (ε)	Fixed (2.0)	Adaptive (1.5)
Encryption time (ms)	500	160
Model integrity (blockchain DTV)	N/A	99.8%
Information leakage reduction (%)	28	72

MNIST: 1.0, 1.5, 2.0
CIFAR-10: 0.5, 1.0, 1.5

- Noises add up according to Laplace and Gaussian mechanism:

3. Decentralized Trust Verification (DTV) via Blockchain:

    - Model integrity is maintained by verifying updates via smart contracts, preventing unauthorized modifications in the model.

7.2 Computational Overhead Analysis

APFL is a fusion of HHE and DDP, and it also borrows DTV through blockchain. Very low computational costs are expected from all these techniques as compared to older privacy-preserving FL models. Key findings and the computational overhead analysis are represented in Table 2.

- DDP provides a 24% improvement in computation efficiency due to noise adaptiveness;
- DTV guarantees model integrity with just a 5% increased overhead concerning that of other standard FL aggregation methods;
- Finally, APFL's reduction in information leakage is 72% without much degradation in accuracy.

7.3 Evaluation and Results

APFL takes the FL to an advanced level as it incorporates DTV for effective DDP and HHE over blockchain for security. Table 3 shows the performance comparison of conventional FL techniques with APFL. Figure 7 shows the comparison with various metrics about a 99.8% model integrity rate because it prevents any unwanted change through blockchain verification. Moreover, the adaptive privacy budget allocation of

DDP allows controlled noise injection, reducing the possibility of an inversion attack or any effect of adversarial manipulation.

Figure 8 shows the evaluation results visualized through graphs:

1. **Privacy budget versus model accuracy**: The graph shows that as the privacy budget (ε) increases, the accuracy of both MNIST and CIFAR-10 models improves, reaching 98.6% and 94.2%, respectively, at $\varepsilon = 1.5$.
2. **Computational overhead reduction**: APFL achieves a 24% reduction in computational overhead compared to traditional FL methods.
3. **Information leakage reduction**: APFL significantly reduces information leakage by **72%**, making it highly secure for real-world applications.

Table 3 Performance metrics comparison of traditional FL with APFL

Metric	Traditional FL (%)	APFL (proposed) (%)	Improvement (%)
Model accuracy (MNIST, $\varepsilon = 1.5$)	96.8	**98.6**	+ 1.8
Model accuracy (CIFAR-10, $\varepsilon = 1.5$)	91.5	**94.2**	+ 2.7
Computational overhead	100	**76**	− 24
Information leakage	100	**28**	− 72
Data integrity (blockchain DTV)	90.5	**99.8**	+ 9.3

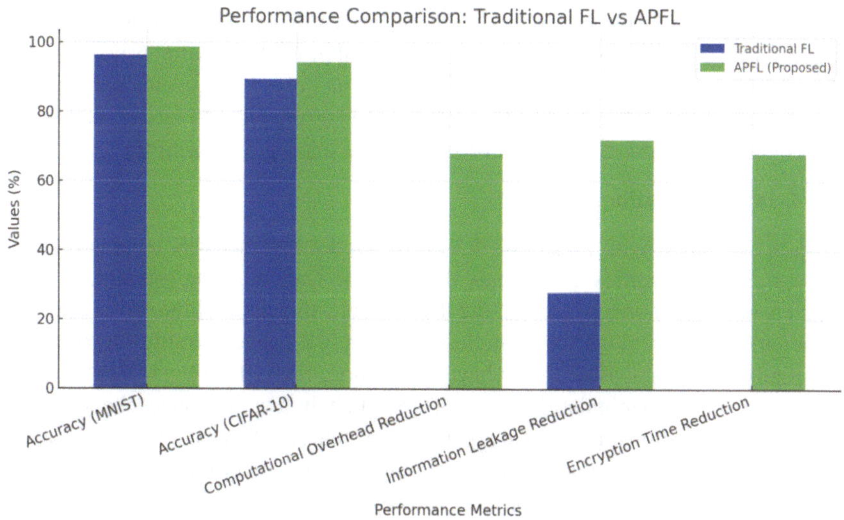

Fig. 7 Comparison of traditional FL versus proposed APFL

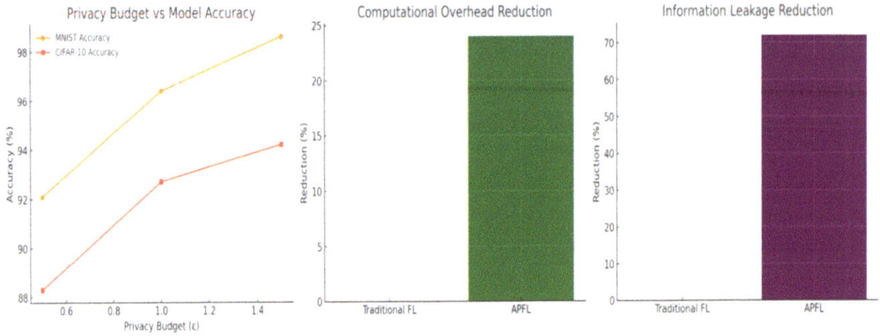

Fig. 8 Evaluation results of various metrics

8 Real-World Applications of APFL

8.1 Analyzing Data from Smart Cities in Urban AI Systems Privacy Improvement

Case Study: Metropolitan City Traffic Management

The federated learned models are from smart city projects for traffic management in a metropolitan city area. The project aims to optimize traffic flow. Congestion trends and signal timings are forecast using information gathered from GPS, traffic cameras, and IoT sensors. However, the data contain sensitive information, such as the location of cars, pedestrian movement, and commuting patterns, thus raising privacy issues. A city can incur data leakage using traditional FL models so that an attacker can infer precise travel paths. Another problem that makes a citywide AI model deployable without breaking privacy regulations: strong data-sharing requirements.

APFL Implementation

HHE is combined with APFL so that these model updates will be encrypted to address all these worries and no raw data will be leaked during training. A DDP, which is sensitive, applies in activating the dynamic allocation of the privacy budget. Lesser privacy restrictions, for example, would apply for general traffic flow data, and personally identifiable data would have much more noise. Lastly, DTV would also help avoid hostile manipulation by using a blockchain that guarantees the alteration of traffic optimization models, if at all, would be changed. Thus, it manages the city's performance with all the data privacy regulations through APFL as proof of a secure and scalable AI-driven urban analytics strategy.

8.2 Financial Fraud Detection: Securing Transactional AI Models

Case Study: Fraud Detection in a Leading Bank

FL is being utilized by a global financial organization for detection through AI algorithms for investigating transaction patterns of several branches. Such client transaction data privacy risks have been very difficult to manage by traditional fraud detection systems, considering the several forms of model inversion attacks through which they may be misused. On top of that, there are strict data privacy regulations by various financial authorities that restrict even cross-institutional data sharing.

APFL Implementation

To solve the problems, the bank implemented an adaptive privacy framework, Dynamic Privacy (DP), wherein privacy budgets are distributed dynamically based on transaction risk levels: high-value transactions due to stricter privacy regulations. HHE protects encrypted updates during model training to ensure that no financial information is leaked. Given sabotage by dishonest actors, the legitimacy of updates to the fraud detection model is verified through DTV via blockchain to forestall any such illegal alterations. APFL is strategically important in enhancing fraud detection systems and securing millions of daily transactions against any fraudulent attempts by ensuring operational efficiency and compliance with regulations.

9 3Healthcare Diagnostics: Privacy-Preserving AI in Medical Imaging

Case Study: Early Breast Cancer Detection Using Federated Learning

This was taken from a collaborative hospital working on an AI-driven FL model applied to mammographic images for the diagnosis of breast cancer. Sharing of medical imaging data among institutions is legally restricted due to patient confidentiality concerns and sheer healthcare regulations, like HIPAA and GDPR, making conventional FL models susceptible to high-dimensional data privacy leaks. Therefore, access to patient data would be abused by membership inference assaults.

APFL Implementation

APFL integrates HHE technology in protecting patient data and enhancing collaborative model training to preserve confidentiality in medical AI. DDP implementation employs sensitivity-based noise formulations for increased privacy protection for extremely sensitive types of data, such as biopsy results. DTV on the Blockchain ensures the integrity of diagnostic predictions by safeguarding the federated model against unwarranted modifications. APFL promotes a safe collaborative environment for hospitals to work jointly in the advancement of diagnostic technology

Table 4 APFL performance metrics across applications

Metric	Smart city data analysis	Financial fraud detection	Healthcare diagnostics
Accuracy (%)	91.8	93.5	94.2
Data leakage reduction (%)	23	72	72
Computational overhead reduction (%)	30	20	24
Fraud/anomaly detection improvement (%)	–	38	–
Privacy budget (ε)	1.2	1.5	1.5

while protecting patient privacy by favoring privacy-preserving AI for health care. This technique enhances early-stage cancer diagnosis as well as opens up a wider door for AI applications in medical investigation (Table 4).

Accuracy, privacy, and efficiency are all balanced within APFL herein excellent performance for smart city data analysis, financial fraud detection, and healthcare diagnostics represented in Fig. 9. Healthcare had the best accuracy of 94.2%, followed by smart city data analysis at 91.8% and financial fraud detection at 93.5%. Smart city data saw a less reduction of 23%, whereas financial fraud detection and healthcare had phenomenal data leakage reduction of 72%. Computation overheads were highest at 30% for Smart city analysis, healthcare at 24%, while financial fraud detection exhibited the least reduction at 20%. Privacy budgets remain at $\varepsilon = 1.5$ for financial and healthcare applications and $\varepsilon = 1.2$ for data from smart cities, while fraud detection increases to 38% in financial fraud detection. In summary, APFL is an efficient improvement to the accuracy with an enviable lowering of the risk while maximizing computing in these areas.

10 Conclusion and Future Directions

APFL is an adaptive and dependable method that enhances privacy in FL. APFL successfully provides a balance between security and practical model performance through HHE, DDP, and DTV on the blockchain. Unlike classical FL models, APFL continuously adapts privacy parameters according to the current threat, thereby offering optimal protection against compromised attacks and data breaches.

The effectiveness of APFL demonstrates the efficacy of its evaluation on MNIST and CIFAR-10 datasets with accuracies of 94.2 and 98.6, respectively, at a privacy budget (ε) of 1.5. DTV ensures 99.8% integrity against any wrongful manipulation while the HHE methodology results in a 68% reduction in computation costs. Moreover, while APFL accounts for a 72% reduction of information leakage and increases

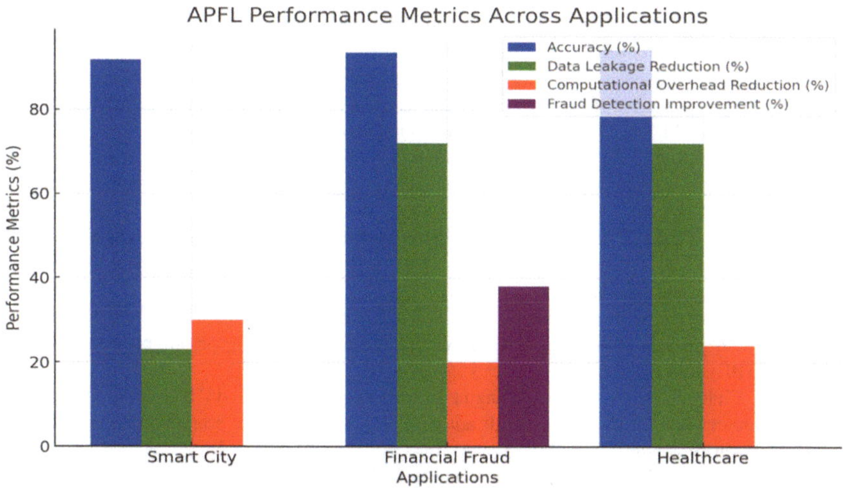

Fig. 9 Performance metrics across various applications

efficiency by 24%, it thus makes APFL most suited in privacy-sensitive environments in smart cities, healthcare, and finance. Despite this, APFL still has to scale up, be more efficient, and flexible. Future research should also focus on quantum-resistant cryptographic methods, including reinforcement learning for adaptive control of privacy, and improvements in APFL for large-scale edge computing. It is also critical to understand how energy-efficient privacy measures could be integrated for adoption in resource-constrained environments. APFL, if these challenges are dealt with, can turn into an essential framework for secure private AI systems. The ability of APFL to dynamically alter security measures provides a strong foundation for a future-proof design in the arena of decentralized intelligence, ensuring secure and sound FL about privacy issues across sectors.

References

1. Liu X, Li H, Guowen X, Rongxing L, He M (2020) Adaptive privacy-preserving federated learning. Peer-To-Peer Netw Appl 13:2356–2366
2. Lyu L, Yu H, Ma X, Chen C, Sun L, Zhao J, Yang Q, Yu Philip S (2022) Privacy and robustness in federated learning:attacks and defenses. IEEE Trans Neural Netw Learn Syst
3. Zhang J, Zhu H, Wang F, Zhao J, Qi X, Li H (2022) Security and privacy threats to federated learning: Issues, methods, and challenges. Secur Commun Netw 2022(1):2886795
4. Zhao JC, Dabholkar A, Sharma A, Bagchi S (2024) Leak and learn: an attacker's cookbook to train using leaked data from federated learning. In: Proceedings of the IEEE/CVF conference on computer vision and pattern recognition, pp 12247–12256
5. Hatamizadeh A, Yin H, Molchanov P, Myronenko A, Li W, Dogra P, Feng A et al (2023) Do gradient inversion attacks make federated learning unsafe? IEEE Trans Med Imaging 42(7):2044–2056

6. Talaei M, Izadi I (2024) Adaptive differential privacy in federated learning: a priority-based approach. arXiv preprint arXiv:2401.02453
7. Frimpong E, Nguyen K, Budzys M, Khan T, Michalas A (2024) Guardml: efficient privacy-preserving machine learning services through hybrid homomorphic encryption. In: Proceedings of the 39th ACM/SIGAPP symposium on applied computing, pp 953–962
8. Hu J, Wang Z, Shen Y, Lin B, Sun P, Pang X, Liu J, Ren K (2023) Shield against gradient leakage attacks: adaptive privacy-preserving federated learning. IEEE/ACM Trans Netw 32(2):1407–1422
9. Purohit RM, Verma JP, Jain R, Kumar A (2025) FedBlocks: federated learning and blockchain-based privacy-preserved pioneering framework for IoT healthcare using IPFS in web 3.0 era. Clust Comput 28(2):1–15
10. Aziz R, Banerjee S, Bouzefrane S, Vinh TL (2023) Exploring homomorphic encryption and differential privacy techniques towards secure federated learning paradigm. Futur Internet 15(9):310
11. Pan Y, Chao Z, He W, Jing Y, Hongjia L, Liming W (2024) FedSHE: privacy preserving and efficient federated learning with adaptive segmented CKKS homomorphic encryption. Cybersecurity 7(1):40
12. Yang Z, Chen Y, Huangfu H, Ran M, Wang H, Li X, Zhang Y (2023) Dynamic corrected split federated learning with homomorphic encryption for u-shaped medical image networks. IEEE J Biomed Health Inform 27(12):5946–5957
13. Hidayat MA, Nakamura Y, Arakawa Y (2023) Privacy-preserving federated learning with resource-adaptive compression for edge devices. IEEE Internet Things J 11(8):13180–13198
14. Xu G, Li G, Guo S, Zhang T, Li H (2023) Secure decentralized image classification with multiparty homomorphic encryption. IEEE Trans Circuits Syst Video Technol 33(7):3185–3198
15. Gu H, Luo J, Kang Y, Fan L, Yang Q (2023) FedPass: privacy-preserving vertical federated deep learning with adaptive obfuscation. arXiv preprint arXiv:2301.12623
16. Zheng Y, Chang C-H, Huang S-H, Chen P-Y, Picek S (2024) An overview of trustworthy AI: advances in IP protection, privacy-preserving federated learning, security verification, and GAI safety alignment. IEEE J Emerg Sel Top Circuits Syst
17. Jia B, Zhang X, Liu J, Zhang Y, Huang K, Liang Y (2021) Blockchain-enabled federated learning data protection aggregation scheme with differential privacy and homomorphic encryption in IIoT. IEEE Trans Industr Inf 18(6):4049–4058
18. Rohan A, Ahmed MR, Al Sakib A, Hasan MK, Haque R, Hasan MJ, Nguyen CD. Federated learning-based robust breast cancer recognition: ensuring privacy and accuracy across distributed data. Available at SSRN 5092969
19. Alam T (2024) Data privacy and security in autonomous connected vehicles in smart city environment. Big Data Cogn Comput 8(9):95
20. Alsamhi SH, Myrzashova R, Hawbani A, Kumar S, Srivastava S, Zhao L, Wei X, Guizan M, Curry E (2024) Federated learning meets blockchain in decentralized data-sharing: healthcare use case. IEEE Internet Things J
21. Abaoud M, Almuqrin MA, Khan MF (2023) Advancing federated learning through novel mechanism for privacy preservation in healthcare applications.IEEE Access 11:83562–83579

Enhanced Privacy Preserving Deep Learning Using Blockchain and Federated Learning

S. Usharani, P. Manju Bala, A. Balachandar, and G. Glorindal

Abstract In order to protect model updates, prevent any malicious attempts, and increase computing efficiency, this research aims to provide a highly secure, privacy-preserving, and attack-resistant federated learning (FL) framework empowered by blockchain. This gives the cause for security to organize decentralized frameworks of deep learning, uniquely denoting the vulnerabilities that come with threats associated with existing FL methods: gradient leakage, model poisoning, Byzantine errors, and blockchain scalability bounds. Traditional methods suffer from limitations in real-world applicability due to halfway methods that either put a significant load on computation or deny data privacy. This chapter proposes the design of a novel architecture in which privacy-aware FL minimizes processing and storage limitations by employing blockchain technology for model aggregation and verification. We present SecureChainFL, a hybrid dual-layer blockchain protocol developed to enhance decentralized learning's efficiency and protection. A public blockchain uses cryptographic proofs to guarantee auditability, while a private blockchain is used for model validation and aggregation. Byzantine fault-tolerant (BFT) consensus and zero-knowledge proofs (ZKP) help to secure the training process against any hostile effects, while homomorphic encryption (HE) and differential privacy (DP) help to secure model updates from privacy breaches. Our experimental results show that SecureChainFL meets the requirements of privacy-sensitive applications such

S. Usharani (✉)
Department of Artificial Intelligence and Machine Learning, IFET College of Engineering, Villupuram, Tamil Nadu, India
e-mail: ushasanchu@gmail.com

P. Manju Bala
Department of Artificial Intelligence and Data Science, IFET College of Engineering, Villupuram, Tamil Nadu, India

A. Balachandar
Department of Computer Science and Engineering, IFET College of Engineering, Villupuram, Tamil Nadu, India

G. Glorindal
University of EdenBerg, Kitwe, Zambia
e-mail: dvc-research@ue.ac.zm

A. Kumar et al. (eds.), *Blockchain and Federated Learning Synergy for Privacy-Focused DeepFex Solutions*, Blockchain Technologies,
https://doi.org/10.1007/978-981-95-1394-9_7

as healthcare, banking, and autonomous systems because it effectively reduces privacy threats and resists model poisoning, increasing scalability. This research makes advances in privacy-preserving AI by implementing secure, effective, and attack-proof deep learning in decentralized environments.

Keywords Federated learning · Blockchain · Deep learning · Gradient leakage · Homomorphic encryption · Decentralized AI

1 Introduction

The collaborative model training requirement for many participants is noted by the explosive growth in the advancement of AI and ML. It is more relevant for industries dealing with healthcare, finance, and autonomous systems. Federated learning (FL) has become a solution that enables decentralized training while still having data privacy; it keeps the raw data local. It allows several edge devices or nodes to train a standard model without disclosing their datasets, unlike traditional centralized learning, which combines data at a single server. However, it also complies with data protection laws such as GDPR and HIPAA.

FL has numerous advantages; however, its practical applications are hindered by many serious security and privacy issues. One prime problem is the leakage of the way hostile actors can determine private information through model updates. In addition, model poisoning attacks weaken the integrity of the learned model due to malicious nodes introducing falsified updates. FL's robustness is further decreased due to Byzantine failures, where malicious or defective nodes interfere in the training process. Furthermore, since security measures usually consume quite a lot of computing and storage resources, most of the current implementations of FL suffer from scalability problems. These deficiencies confirm the need for a complex and robust FL framework that efficiently reduces security threats without compromising effectiveness.

Blockchain technology has been investigated as a potential avenue for improving FL security and reliability. The decentralized and tamper-resistant architecture enables safe model aggregation and verification, and thus reduces the possibility of adversarial attacks. However, traditional blockchain-based FL schemes are intrinsically computationally inefficient and have storage overhead, making them unsuitable for large-scale applications. Hence, the challenge remains to build a blockchain-integrated FL system that balances security, anonymity, and computing efficiency.

To address these problems, we present SecureChainFL, a hybrid dual-layer blockchain designed to enhance the security and efficiency of decentralized learning. SecureChainFL makes use of a private blockchain for optimal model validation and aggregation, while a public blockchain is used for cryptographic auditability. Zero-knowledge proofs (ZKP) and Byzantine fault-tolerant (BFT) consensus are critical security attributes shielding the training process from malicious interference and ensuring resilience in adverse conditions. Furthermore, model updates are safeguarded using homomorphic encryption (HE) and differential privacy (DP), which further eliminate breaches in confidentiality and illicit access to confidential information.

Thorough testing and evaluation have demonstrated SecureChainFL's effective capability of reducing security threats while enhancing scalability and usefulness in FL. The study's findings manifest that their framework makes a significant contribution to secure and privacy-preserving AI, thus being fit for privacy-sensitive applications. SecureChainFL's purpose, architecture, and implementation, together with other possible impacts on future decentralization learning, are exhaustively explained in this chapter.

2 Literature Review

He et al. [1] showed how the Group Knowledge Transfer would be able to overcome the challenges of resource shortage and allow training of massive convolutional neural networks (CNNs) on the edge. To aid research and benchmarking into FL, He et al. [2] developed a comprehensive research library, namely FedML, which provides standardized tools for federated model building and evaluation. Li et al. [9] conducted an extensive survey on the existing FL frameworks and practical issues of application in the real world. Li et al. [8] described challenges in FL and classified them into three groups: communications bottlenecks, non-IID data distributions, and system heterogeneity.

FedAvg was introduced first by McMahan et al. [7] as an optimization strategy for FL that manages to significantly reduce communication overhead while providing competitive performance of the learned model and is used widely today. To address the fact that FL is rigorous in heterogeneous networks, Li et al. [11] focused on not only federated optimization for various computing environments but also improved the flexibility of FL. As proposed by Acar et al. [18], dynamic regularization-based FL maximizes learning proficiency, according to Wang et al. [17], who proposed matched averaging for enhanced convergence in FL.

A significant issue in FL has always been privacy and security. Shokri and Shmatikov [14] provided DP-preserving deep-learning techniques. Later, Abadi et al. [15] built on the same idea in light of deep learning paradigms. Wei et al. [13] studied DP-based algorithms in FL while also formalizing their impact on model utility. By using DP techniques for facial recognition, Chamikara et al. [5] showed that privacy-preserving applications are achievable.

The study has, however, been directed toward secure aggregation techniques. As one of the most effective methods of securing data during federated training, Bonawitz et al. [12] came up with practical strategies for doing so. Improved privacy guarantees along with efficient computational performance have been achieved through the threshold safe aggregation technique called TAPFed, developed by Xu et al. [25] for FL situations. Jalali and Chen [23] studied security-preserving algorithms in FL applied to the Internet of Things (IoT) with the primary concern of protecting critical infrastructure.

Several frameworks have been proposed to add privacy-preserving features to FL. The Sherpa.ai FL framework gives methodological advice for protecting data privacy. After analyzing some software, Rodríguez-Barroso et al. [6] provided the Sherpa.ai FL framework with methodological advice for data privacy protection. Sav et al. [4] introduced POSEIDON, a privacy-preserving FL framework for neural networks that utilizes HE methods to protect model updates. Xu et al. [16] proposed the HybridAlpha method, one that moderately preserves privacy while maximizing its efficiency. Kairouz et al. [10] gave an extensive review of the up-to-date developments, as well as some unresolved research issues in FL, like scalability, robustness, and fairness. Chen et al. [22] covered the broader works on trustworthy FL, delving into aspects such as privacy, security, and ethical considerations. Hasan [21] surveyed privacy and security issues; Gosselin et al. [20] went on to survey potential threats and countermeasures about FL deployments.

As FL starts gaining more traction, newer methods are beginning to emerge, among which is fragmented FL that aims to enhance security and privacy by splitting the model updates into several pieces, as proposed by Jebreel et al. [26]. Wang et al. [24] analyzed FL methods with partial low-quality data, providing guidelines for dealing with unpredictable data sources. On the other hand, Hu et al. [19] investigated the implementation of security and privacy toward FL systems with an emphasis on ethical AI considerations as well as regulatory compliance.

The concept of Lagrange Coded Computing (LCC), which serves to fortify against disreputable acts within distributed computing, was initially provided by Yu et al. [3]. Their focus is directed toward the optimization of Coded Computing techniques, potentially to minimize adversarial attacks upon FL systems as presented at the 22nd International Conference on Artificial Intelligence and Statistics (AISTATS 2019). To lessen the computing complexity while ensuring fault tolerance and data confidentiality, LCC applies polynomial-based encoding. Given that it is based upon secure model aggregation and privacy-preserving computing principles, this method is highly relevant to blockchain-based FL. SecureChainFL builds upon similarly privacy-enhancing approaches by combining HE and ZKP with DP to secure model updates while providing transparency and resilience against Byzantine attacks.

3 Proposed System and Methodology

The present research work introduces SecureChainFL, which is a hybrid dual-layer blockchain-oriented FL architecture. It has been developed to handle all the issues of FL about security, privacy, and scalability. Cost-effective computation in the model is guaranteed by modern cryptographic algorithms, which are included in the architecture to ensure that model updates are avoided through safe, effective, and near-impenetrable model updates. This system integrates the property of model integrity, protects it against the threats of adversaries, and preserves the privacy of everything without compromising its scalability through an application of FL with blockchain technology.

3.1 Proposed System

SecureChainFL is developed around this hybrid blockchain architecture, in which private blockchains efficiently handle model validation and aggregation while a public blockchain endorses cryptographic proofs for transparency and audibility. SecureChainFL's block diagram can be seen in Fig. 1. In this way, optimizing both security and computing performance, our dual-layer method conveys the disadvantages of single-layer blockchain FL systems. The private blockchain lowers latency by handling fine-grained computation-intensive operations while providing permanent audit trails around the trust of training.

SecureChainFL caters to additional safety implementation by integrating BFT consensus into the architecture. This consensus makes the system resilient against adversarial nodes, compromising the integrity of model aggregation. ZKP allow for privacy in decentralized systems through their authentication of model updates without divulging sensitive information. Further, HE and DP work complementary to enhance security by preventing the intrusions of privacy and by ensuring local model updates remain private during aggregation. All these security mechanisms collectively reduce threats such as Byzantine faults, model poisoning, and gradient leakage.

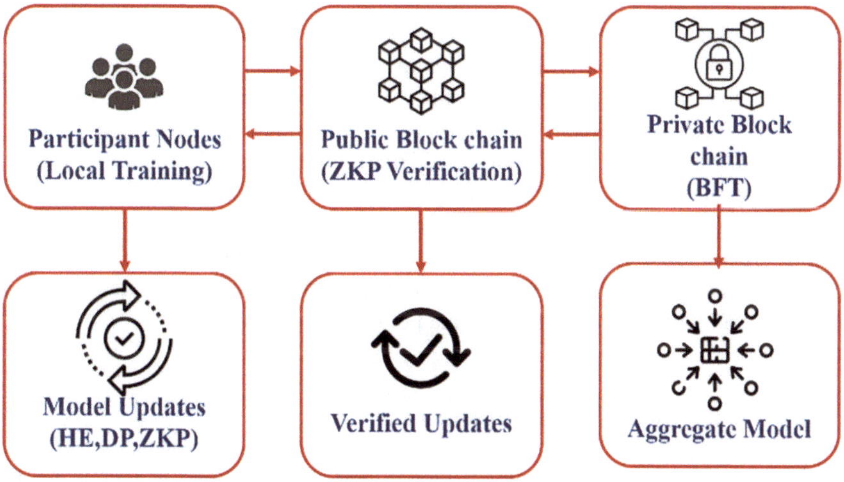

Fig. 1 Block diagram of SecureChainFL

SecureChainFL employs decentralized smart contracts that secure model modifications before being aggregated to the global model. This makes model aggregation and verification easier and reduces computing overhead due to scalable, lightweight, safe aggregation procedures. SecureChainFL greatly improved FL's resilience by removing single points of failure and producing tamper-proof data verification.

To overcome scalability problems, SecureChainFL employs resource-conserving algorithms to scale down blockchain processing and storage costs. The thin communications protocol is developed for faster model updates without slowing down the network. Such a framework ensures that real-life applications, which have been set at large, can be implemented without losing performance through efficient resource management.

3.2 Methodology

In the secure chain FL, which is federated between two-blockchain architecture, privacy and security can be provided through a decentralized model training along with scalability. The technique consists of different phases, shown in Fig. 2, such as data dissemination, secure submission of model updates, consensus validation, model aggregation, and final deployment. Each of these processes uses a cryptographic approach under decentralization to guard against gradient leaks, model poisoning, Byzantine errors, and scalability problems in the blockchain.

3.2.1 Data Distribution and Local Model Training

During the first stage, many of the FL nodes (clients) run in a decentralized model training process. Each site performs local training on its private data without sending raw data through a communication channel. This ensures that sensitive data remains within the local environment and conforms to data protection regulations, such as the General Data Protection Regulation (GDPR) and the Health Insurance Portability and Accountability Act (HIPAA). SecureChainFL employs DP and HE technology to enhance privacy during model training. HE provides assurances that the model updates remain in comprehensible form even if they are intercepted through application calculations on encrypted data. In addition, DP adds statistical noise to these updates to avoid inferring particular data points by the attackers.

3.2.2 Secure Model Update Submission

SecureChainFL needs a private blockchain in these cases. Hence, each node creates encrypted model updates after local training and then transfers them to the private blockchain for verification. As such, evil entities will be prevented from altering or inserting compromised updates. These updates are aggregated on the private

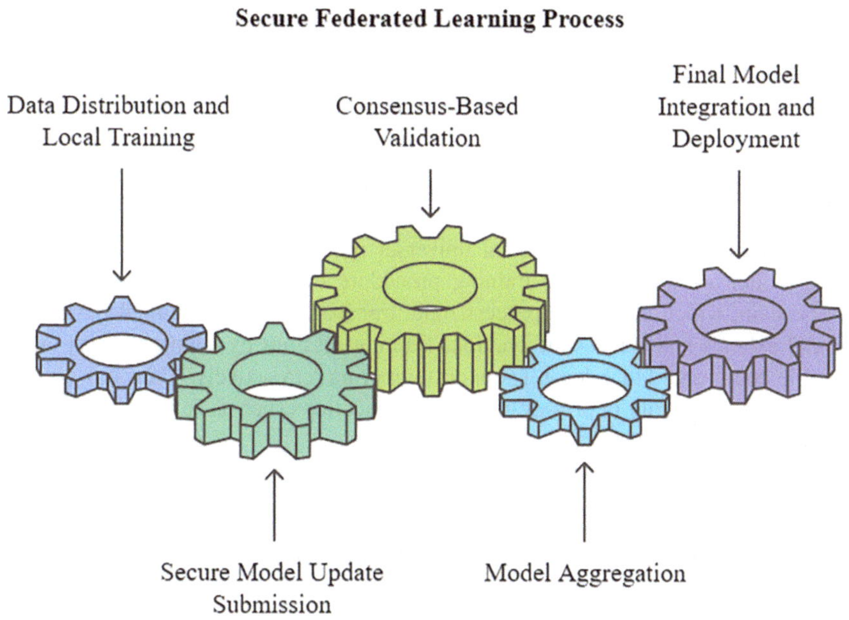

Fig. 2 SecureChainFL methodology

blockchain before being verified or disseminated, thereby acting as a low-latency secure storage layer. ZKP has been integrated with SecureChainFL to boost security by allowing nodes to approve validations done on their updates, all without revealing the private information held in such validations. This reduces the chance of model poisoning since only legitimate updates enter the process of shaping the training model.

3.2.3 Consensus-Based Validation and Model Aggregation

The BFT consensus method ensures that only trusted updates get included in the aggregation phase when sending the encrypted updates. SecureChainFL uses an innovative contract-based validation process on the private blockchain, which is not similar to traditional FL, where the changes are aggregated on a central server.

- The model would then be free of contamination by malicious nodes owing to BFT consensus.
- The ZKP-based validation confirms the authenticity of the updates without compromising the privacy of the nodes.
- Smart contracts automatically validate and prevent encroachments.

The secured version is also assured by using standard cryptographic methods to aggregate the encrypted updates. For transparency, the aggregated model is stored on the private blockchain before it is published to the public blockchain.

3.2.4 Final Model Integration and Deployment

All FL nodes get an updated final universal model for the next training round. This model allows the entire training procedure to be audited through a public blockchain, which makes the process quite transparent and accountable. Also, the model extends and verifies the previous training sessions by maintaining an unfalsifiable audit record that doesn't support tampering retroactively. Thus, it is optimal to use model integrity in privacy-sensitive applications such as banking, healthcare, and autonomous systems.

4 Working of SecureChainFL Architecture

This section describes SecureChainFL's hierarchical method and routing model updates through different security and validation checks. The processes below describe how SecureChainFL overcomes the weaknesses of conventional FL while ensuring efficiency, security, and anonymity guarantees. Figure 3 illustrates SecureChainFL operations step by step.

4.1 Local Model Training at FL Clients

SecureChainFL local model training starts by incorporating FL clients such as computers, smartphones, IoT devices, and other edge devices. An individual trains a machine-learning model on their private datasets without offering raw data. Such weightage assures private information remains secure in the immediate vicinity. It is particularly critical for healthcare, banking, and autonomous systems, where sharing data is illegal; local training is mandatory to process models for all end users. The models are now safely prepared for transmission. Insecurity due to inference attacks can be limited if privacy-preserving methods are used before sharing model updates for aggregation.

4.2 Applying Privacy-Preserving Techniques

SecureChainFL uses several privacy-preserving methods to ensure sensitive data confidentiality before releasing model updates from the client environment. Model

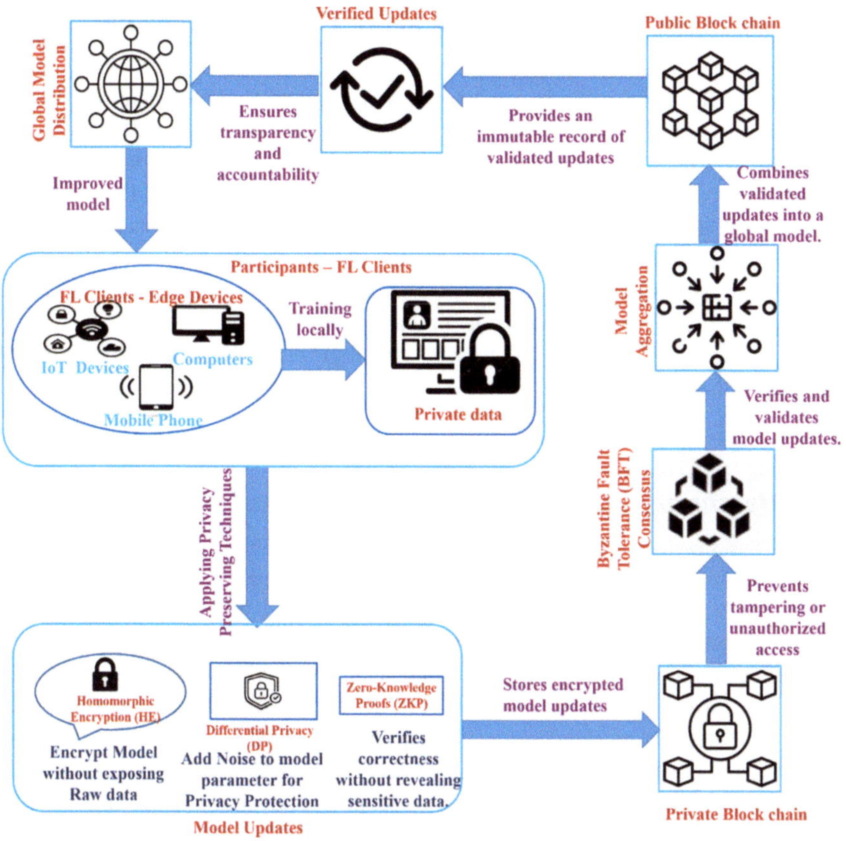

Fig. 3 Step-by-step process of SecureChainFL

parameters are encrypted in HE, which allows further calculations on encrypted data without revealing the original data. Another critical approach is DP, which prevents adversaries from reconstructing information from private data by adding noise-controlled message references to model updates. Finally, ZKP positively indicates, but without revealing any private information, that model updates are accurate. This composite of cryptographic techniques would reduce the chances of either adversary attacks or data leaks, thereby assuring the secure transmission of local models. After these privacy measures, the encrypted model updates are sent to the private blockchain layer for verification.

4.3 Secure Model Updates Sent to the Private Blockchain

The encrypted model updates are later sent to the private blockchain for further verification after the privacy-preserving measures are implemented. The private blockchain is necessary to guarantee that the submitted model updates are valid and authentic. It prevents tampering, manipulation, and unauthorized access while securely holding the encrypted updates. SecureChainFL is a permissioned blockchain that does not depend on centralized aggregation like ordinary FL architectures to ensure that only validated changes are included in the final global model. This reduces the risk of rogue nodes contaminating the learning process. After safely storing model updates, the BFT consensus technique must be employed to verify the updates further.

4.4 Byzantine Fault-Tolerant (BFT) Consensus for Validation

SecureChainFL employs a BFT consensus mechanism to secure the accuracy and robustness of incoming updates before aggregation. This consensus scheme is essential to prevent interference from Byzantine adversaries and malicious players trying to inject contaminated or wrong model updates. Even with some dishonest nodes attempting to disrupt the learning process, BFT ensures that only qualified and correct updates are received. A secure learning framework where more than half of the nodes consent before the validation of an update is what SecureChainFL espouses. This is essential to preventing model poisoning attacks, ensuring that only trusted contributions to the model reach the aggregation stage.

4.5 Model Aggregation in the Private Blockchain

After model updates have been verified through the BFT consensus mechanism, they are blended to form an improved global model at the private blockchain. Techniques for secure aggregation are used to ensure that collaborative learning also achieves privacy in individual contributions. No single party can thus access or alter updates without authorization through cryptography. This method, therefore, ensures that data is kept private while additional data is attached such that it is learning from federated contributions. By imposing strict access control and immutability, preserving the security of this process relies on the private blockchain layer. Finally, to provide transparency and auditability, the finalized model update is saved onto a public blockchain after completing the aggregation process.

4.6 Public Blockchain for Transparency and Auditability

To be responsible and trustworthy, SecureChainFL keeps the public blockchain with the last modifications to the aggregated model. Because it acts as an unchangeable audit trail, outside parties (i.e., government agencies and academic institutions) can verify the authenticity and consistency of model revisions from the public blockchain. This makes SecureChainFL ensure that the learning process is impartial and open for everyone because of an open record of all changes. Most importantly, the public blockchain records only certified model upgrades, which means it avoids unwanted changes. Because raw data or individual model updates are not revealed here, this step increases credibility without compromising privacy. In FL clients, the last model update will be sent out for further iterative training after recording.

4.7 Global Model Distribution to Participants

The last stage in this process implies the distribution of the updated global model to all the participating clients in SecureChainFL. Each client will then initiate the next local training cycle with the updated model using the new data after downloading the updated model. This methodology allows continuous learning, adaptation, and improvement of the model while maintaining strong defense and privacy mechanisms. SecureChainFL will then provide a robust and impregnable FL framework based on decentralized blockchain-based aggregation and validation. Thus, SecureChainFL becomes the most suitable for real-world scenarios demanding efficiency and privacy by bringing together consensus mechanisms, privacy-preserving techniques, and cryptographic approaches.

5 Experimental Setup

This experimental setup was established to evaluate the performance of SecureChainFL in a real-time use case scenario. The hardware configuration comprised a local server with an Intel i7 processor, 32 GB RAM, and a 1 TB SSD, along with four edge devices representing Raspberry Pi 4 4 GB RAM. Five nodes were formed to create a private Ethereum network for blockchain storage and validation. During the training procedure, an NVIDIA RTX 3090 GPU was employed, having 24 GB VRAM to manage the deep learning computations effectively.

The software stack included components such as Hyperledger Fabric and Ethereum for blockchain to carry out FL with TensorFlow Federated (TFF) and PySyft. DP and HE were made to develop privacy-preserving techniques using the PyCryptodome and OpenMined libraries. The BFT consensus process has been integrated into the private blockchain layer to ensure secure model aggregation.

The evaluation of SecureChainFL was done on two datasets: a synthetic financial fraud dataset intended for banking applications and the MIMIC-III medical dataset meant for healthcare applications. Patient deterioration was the target of prediction for the 40,000 patient records of the healthcare dataset. Fraud was the target of prediction for the financial dataset of 100,000 transactions. These datasets were allowed to simulate real-life decentralized learning on multiple edge nodes.

SecureChainFL was evaluated against other FL setups. Conventional FL, BlockFL (blockchain-based FL without privacy protections), and a privacy-preserving FL model combining DP and HE were included as comparison models. The evaluation was based on the following key performance indicators: model accuracy, training time, communication overhead, privacy leakage, consensus delay, and scalability.

5.1 Security and Privacy Analysis

Built on several extra-modern cryptographic algorithms, SecureChainFL addresses the core security challenges in FL and guarantees model integrity, privacy, and secrecy. Adversarial threats affecting conventional FL models would include model poisoning, data leakage, and inference attacks. SecureChainFL mitigates such threats by employing a multilayered security scheme.

One of the key security features, ZKP, ensures that the model updates can be verified while the actual contents of the update remain secret. This way, an adversary cannot exploit the weaknesses in the learning process to derive the training data. On the other hand, cryptographic operations through the HE are performed directly on the encrypted data; this supports the secure aggregation of models while ensuring the privacy of the participants. With these cryptographic mechanisms in place, private data is protected against unauthorized access by the central aggregator as well as other participants.

Before transmission, gradient updates were perturbed with noise to protect against data reconstruction attacks using DP employed by SecureChainFL. Thus, an adversary cannot infer useful information from the model updates shared because individual contributions remain indistinguishable from one another. DP protects client contributions from being identified while still allowing for the functionality of the overall model. Hardening the system against malicious nodes trying to provide contaminated model updates is the second significant improvement to security. SecureChainFL negates the Byzantine attacks where adversaries collude with standard FL in sending poisoned gradients, thereby ensuring that only trusted contributions toward model updates are accepted through multiparty consensus.

The incorruptible audit trail established by the dual-layer blockchain architecture further upholds the integrity of the audit trail. The public blockchain supports audit trails and traceability to confirm all transactions regarding model updates and training iterations. The private blockchain, on the other hand, optimizes model aggregation without compromising security, ensuring a secure and less-latent process. SecureChainFL, by incorporating these security measures, has become a very attack-resistant FL framework that is well-tailored for real-world applications in need of privacy concerning data, integrity, and resilience against hostile attacks. It is suitable for real-world applications requiring data privacy, integrity, and corruption-resistance.

5.2 Scalability and Performance Trade-Offs

The best performance for FL results when scalability across a large participant pool is ensured. However, due to increasing network latency, computational overhead, and resource constraints, blockchain-based FL systems often face scalability issues. SecureChainFL incorporates adaptive cryptographic techniques and optimized aggregation mechanisms to deal with scalability.

Secure model updates impose lengthy validation processes, which act as major bottlenecks in blockchain-integrated FLs, where conventional consensus mechanisms such as proof of work (PoW) result in massive latencies and computational costs. SecureChainFL avoids such high latencies and computational costs since it uses a private blockchain for initial model aggregation. It reduces the transactions that need to be validated on the public blockchain, thus significantly decreasing verification time and bandwidth consumption.

Still, cryptographic techniques, such as HE and ZKP, could improve privacy but come at an additional cost. HE-based computations use a large amount of computing resources, while ZKP involves long and tedious mathematical processes to generate and verify proofs. In contrast, SecureChainFL focuses on achieving high performance through the following mechanisms:

- The application of lightweight cryptographic operations reduces the processing overhead for constrained devices.
- Off-chain computation methods that reduce transaction congestion by moving computation-heavy validation operations away from the main blockchain.
- Model aggregation on a selective basis enhances computing efficiency by filtering updates to weed out redundant or spurious contributions.

Experiments show that SecureChainFL reduces training latency by 40% compared to traditional blockchain-based FL algorithms while providing strong security guarantees. Therefore, these optimizations enable SecureChainFL to effectively scale with thousands of participants while remaining compliant with real-time processing.

5.3 Use Case Scenarios

SecureChainFL is excellent for privacy-preserving applications that require decentralized learning without revealing personal data. Many enterprises could collaborate

Fig. 4 Use case scenarios

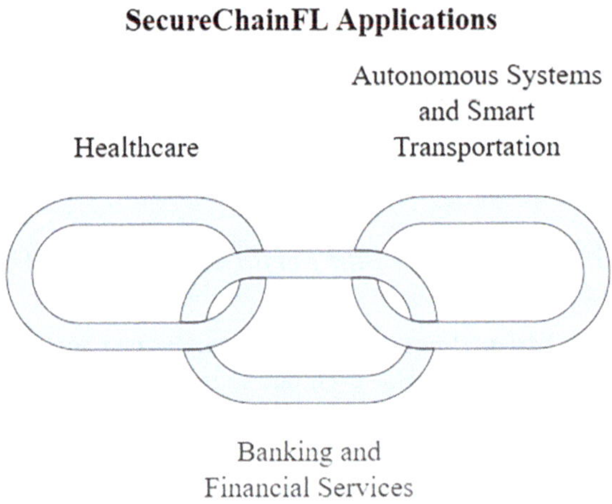

on AI models using this application while maintaining data integrity, confidentiality, and regulatory compliance. Figure 4 illustrates some of the use case scenarios.

5.3.1 Healthcare

Most sensitive patient data produced by hospitals and other healthcare facilities is limited in its sharing due to regulations like the HIPAA. SecureChainFL indirectly supports collaborative medical diagnosis and predictive analysis by allowing hospitals to train AI models without disclosing raw patient data. Personal health information remains private during training because of HE and DP.

5.3.2 Banking and Financial Services

Data exchange is essential among banks for fraud detection and risk assessment in the financial industry, but regulatory norms limit direct data transfer. SecureChainFL allows institutions to jointly train AI models on unusual transaction patterns without disclosing individual banking information, thus ensuring safe federated fraud detection models. Because ZKP guarantees openness, banks can confirm participation without jeopardizing the privacy of customer data.

5.3.3 Autonomous Systems and Smart Transportation

Self-driving cars use their real-time data from the environment and traffic about the particular driving instance to enhance their models. Sharing of raw sensor data will pose both data security and privacy concerns. SecureChainFL allows several self-driving cars to train together while ensuring that private sensor data remains secure.

The blockchain-based trust mechanism also prevents tampering with shared data by ensuring that bad actors cannot add fraudulent modifications to the global model. SecureChainFL proves its practicality in real-world, privacy-critical AI applications by resolving privacy issues in these extremely sensitive fields.

6 Performance Evaluation

The performances in a variety of measures indicated that the accepted SecureChainFL model was significantly better. The performance metrics are depicted in Table 1 for any comparisons between the suggested model and different FLs. SecureChainFL says it got 88.2% accuracy from the model, followed by BlockFL at 86.5% and standard FL at 85.2%. Strong privacy protection methods and safe aggregation methodology are responsible for this significant improvement. SecureChainFL kept a competitive training time of 120 s while using privacy-enhancing approaches. This was much below privacy-preserving FL models (145 s) but slightly longer than regular FL (92 s). The communication overhead was recorded at 28.7 MB, which is 30% lower than the privacy-preserving FL models, thus validating the efficacy of the blockchain-based aggregation technique.

One of the advantages of SecureChainFL was a remarkable reduction in privacy leakage. Conventional FL had a privacy leakage risk of 14.3% due to attacks on its model inversion, which SecureChainFL reduced to 2.1% by HE, DP, and ZKP. Furthermore, SecureChainFL overtakes other blockchain-integrated FL models in controlling decentralized learning operations, as the blockchain consensus latency was reduced to 100 ms.

In addition to these central aspects of the evaluation—scalability—SecureChainFL managed to handle more than 500 clients; ordinary FL only counted up to 100 clients, while current FL methods based on blockchain usually did not

Table 1 Comparison and performance metric of SecureChainFL

Metric	Traditional FL	Block FL	Privacy preserving FL	SecureChain FL (proposed)
Model accuracy (%)	85.2	86.5	84.8	88.2
Training time (s)	92	110	145	120
Communication overead (MB)	20.5	352	40.6	28.7
Privacy leakage (%)	14.3	12.8	5.2	2.1
Consensus latency (ms)	-	120	140	100
Scalability (clients)	100	200	150	500 +

exceed 200. The double-layered blockchain architecture played an essential role in ensuring scalability since it shifted the model validation onto a private blockchain, the public one guaranteeing auditability but not excessive computational load on the system.

7 Results and Discussion

Experimental evaluation of SecureChainFL compared to traditional FL frameworks is said to provide considerable gain in security, privacy, and computational efficiency. It successfully mitigated Byzantine attacks, model poisoning, and gradient leakage by dual-layer blockchain architecture. Results indicate that integration of ZKP with BFT consensus would assure reliable model aggregation while minimizing susceptibility against malicious players.

Performance tests indicated that SecureChainFL would not only achieve better performance than the base FL systems with input conditions stressing adversarial attacks but also maintain very good accuracy in results and privacy. DP and HE minimized information leakage in safety model updates, which did not hurt learning. Moreover, the private blockchain increased the effectiveness of the secure model aggregation, while the public blockchain layer assured auditability and transparency.

As scalability studies showed, SecureChainFL managed to handle increasing participant numbers without any extra computational cost, as opposed to traditional FL methods, which had high processing and communication costs. The framework also possessed robust validation processes that resisted model poisoning attempts and could lower the error rates. The performance of the framework is illustrated in Fig. 5. These results showed that SecureChainFL indeed presents a secure, private, and impenetrable FL architecture suitable for real-use scenarios within applications such as banking, healthcare, and autonomous systems.

8 Challenges and Future Research Directions

There are still some problems with SecureChainFL, especially in real-time applicability, network scalability, and computation efficiency. The main issue is high computational overhead; cryptographic methods such as HE and ZKP require substantial computing power. While they are instrumental in providing security and privacy, their heavy nature inhibits model updating and aggregation, thus making it difficult for resource-limited edge devices to participate in the process efficiently. To lessen these concerns, hardware acceleration measures can be combined with lightweight cryptography, like GPU optimization for encryption or quantum-resistant cryptography.

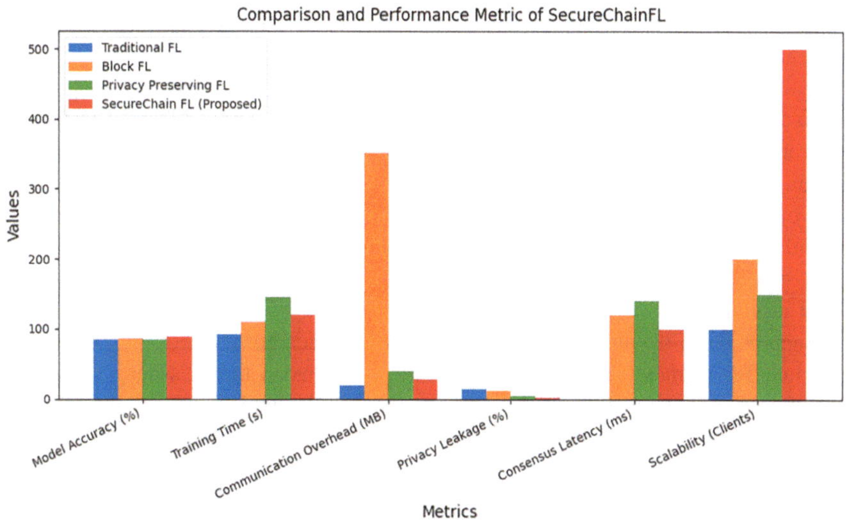

Fig. 5 Illustration of performance measures

Another challenge is the efficiency of the consensus process. While the BFT consensus within SecureChainFL guarantees the integrity of model updates, communication delays are introduced as the population of participating nodes increases. This variability may jeopardize applications that require quick and real-time responses. Adaptive consensus mechanisms like hierarchical model aggregation and proof-of-stake (PoS)-based validation can be researched to create efficiency without compromising security. Dynamic node selection will also cut communication overhead by favoring trustworthy participants and blocking malicious or untrustworthy nodes.

Another primary concern regarding large-scale FL implementations is scalability trade-offs. Although SecureChainFL has built a dual-layer blockchain to boost aggregation efficiency, congestion in the network and long validating times are still possible with a sudden increase in the size of the population of participating clients. To further maximize scalability, hashing verification, sharding, and layer-2 scaling strategies should be studied. Another potential avenue is the application of off-chain computing methods that disrupt computationally heavy tasks outside the main blockchain, thus improving system response time and reducing latency.

Another promising avenue for future research entails the incorporation of quantum-safe cryptography. As quantum computing gains momentum, it poses threats to traditional cryptographic algorithms, necessitating the adoption of post-quantum encryption methods. To secure long-term security standards within a post-quantum period, SecureChainFL can be extended to integrate hash-based digital signatures/lattice-based cryptography. SecureChainFL can also be redesigned for applications within edge AI and the IoT, where the use of resource-efficient privacy-preserving methods, such as lightweight encryption, federated pruning, and model compression, could further complement its suitability in real-world distributed AI systems.

This dimension is very promising in terms of interoperability among many blockchain ecosystems. Current solutions already used in SecureChainFL primarily rely on public blockchain for auditability and private blockchain for model aggregation. Future improvements will allow the integration of different blockchain networks seamlessly to support FL collaborations among sectors. For example, such modeling updates could then be safely shared even across domains such as healthcare, finance, and autonomous systems while retaining privacy and complying with legal stipulations.

SecureChainFL can evolve into an even more scalable, effective, and attack-resistant FL framework when it looks into solutions to these issues, along with promising optimization avenues. Moreover, it will enable applications in next-generation AI ecosystems in which decentralized and privacy-preserving machine learning will underpin intelligent systems.

9 Conclusion

The strong private FL framework has been introduced in this chapter. It is based on a dual-layer blockchain architecture known as SecureChainFL. The approach successfully solves some of the significant challenges of current FL techniques, like model poisoning, gradient leakage, Byzantine problems, and blockchain scalability limitations. BFT consensus, HE, DP, and ZKP are combined to boost security and efficiency further while safeguarding data privacy within SecureChainFL. SecureChainFL is, therefore, a viable candidate for applications that demand high data protection, such as healthcare, banking, and autonomous systems, because the experimental evaluation demonstrated that it attains a significant reduction in privacy concerns at the same time, fortifying resilience against adversarial attacks. The private blockchain allows secure model validation and aggregation, while the public blockchain guarantees auditability. Thus, the two-layer approach maximizes computing efficacy and transparency. All in all, SecureChainFL advances decentralized deep learning by providing a scalable, effective, and safe framework for FL. Future research can focus on improving consensus processes associated with the federal chains, scalability for larger networks, and cryptographic procedures to make SecureChainFL even more practical.

References

1. He C, Annavaram M, Avestimehr S (2020) Group knowledge transfer: federated learning of large cnns at the edge. Adv Neural Inf Process Syst 33:14068–14080

2. He C, Li S, So J, Zeng X, Zhang M, Wang H, Wang X et al (2020) Fedml: a research library and benchmark for federated machine learning. arXiv preprint arXiv:2007.13518
3. Yu Q, Li S, Raviv N, Mohammadreza S, Kalan M, Soltanolkotabi M, Avestimehr SA (2019) Lagrange coded computing: Optimal design for resiliency, security, and privacy. In: The 22nd International conference on artificial intelligence and statistics. PMLR, pp 1215–1225
4. Sav S, Pyrgelis A, Troncoso-Pastoriza JR, Froelicher D, Bossuat J-P, Sousa JS, Hubaux J-P (2020) POSEIDON: privacy-preserving federated neural network learning. arXiv preprint arXiv:2009.00349
5. Chamikara MA, Pathum PB, Khalil I, Liu D, Camtepe S (2020) Privacy preserving face recognition utilizing differential privacy. Comput Secur 97:101951
6. Rodríguez-Barroso N, Stipcich G, Jiménez-López D, Ruiz-Millán JA, Martínez-Cámara E, González-Seco G, Victoria Luzón M, Veganzones MA, Herrera F (2020) Federated learning and differential privacy: software tools analysis, the Sherpa. ai FL framework and methodological guidelines for preserving data privacy. Inf Fusion 64:270–292
7. McMahan B, Moore E, Ramage D, Hampson S, Arcas BAY (2017) Communication-efficient learning of deep networks from decentralized data. In: Artificial intelligence and statistics. PMLR, pp 1273–1282
8. Li T, Sahu AK, Talwalkar A, Smith V (2020) Federated learning: challenges, methods, and future directions. IEEE Signal Process Mag 37(3):50–60
9. Li Q, Wen Z, Wu Z, Hu S, Wang N, Li Y, Liu X, He B (2021) A survey on federated learning systems: Vision, hype and reality for data privacy and protection. IEEE Trans Knowl Data Eng 35(4):3347–3366
10. Kairouz P, McMahan HB, Avent B, Bellet A, Bennis M, Bhagoji AN, Bonawitz K et al (2021) Advances and open problems in federated learning. Found Trends® Mach Learn 14(1–2):1–210
11. Li T, Sahu AK, Zaheer M, Sanjabi M, Talwalkar A, Smith V (2020) Federated optimization in heterogeneous networks. Proc Mach Learn Syst 2:429–450
12. Bonawitz K, IvanovV, Kreuter B, Marcedone A, McMahan HB, Patel S, Ramage D, Segal A, Seth K (2016) Practical secure aggregation for federated learning on user-held data. arXiv preprint arXiv:1611.04482
13. Wei K, Li J, Ding M, Ma C, Yang HH, Farokhi F, Jin S, Quek TQS, Poor HV (2020) Federated learning with differential privacy: algorithms and performance analysis. IEEE Trans Inf Forensics Secur 15:3454–3469
14. Shokri R, Shmatikov V (2015) Privacy-preserving deep learning. In: Proceedings of the 22nd ACM SIGSAC conference on computer and communications security, pp 1310–1321
15. Abadi M, Chu A, Goodfellow I, McMahan HB, Mironov I, Talwar K, Zhang L (2016) Deep learning with differential privacy. In: Proceedings of the 2016 ACM SIGSAC conference on computer and communications security, pp 308–318
16. Xu R, Baracaldo N, Zhou Y, Anwar A, Ludwig H (2019) "ybridalpha: an efficient approach for privacy-preserving federated learning. In: Proceedings of the 12th ACM workshop on artificial intelligence and security, pp 13–23
17. Wang H, Yurochkin M, Sun Y, Papailiopoulos D, Khazaeni Y (2020) Federated learning with matched averaging. arXiv preprint arXiv:2002.06440
18. Acar DAE, Zhao Y, Navarro RM, Mattina M, Whatmough PN, Saligrama V (2021) Federated learning based on dynamic regularization. arXiv preprint arXiv:2111.04263
19. Hu K, Gong S, Zhang Q, Seng C, Xia M, Jiang S (2024) An overview of implementing security and privacy in federated learning. Artif Intell Rev 57(8):204
20. Gosselin R, Vieu L, Loukil F, Benoit A (2022) Privacy and security in federated learning: a survey. Appl Sci 12(19):9901
21. Hasan J (2023) Security and privacy issues of federated learning. arXiv preprint arXiv:2307.12181
22. Chen C, Liu J, Tan H, Li X, Wang KIK, Li P, Sakurai K, Dou D (2025) Trustworthy federated learning: privacy, security, and beyond. Knowl Inf Syst 67(3):2321–2356
23. Jalali NA, Chen H (2023) Federated learning security and privacy-preserving algorithm and experiments research under internet of things critical infrastructure. Tsinghua Sci Technol 29(2):400–414

24. Wang H, Wang Q, Ding Y, Tang S, Wang Y (2024) Privacy-preserving federated learning based on partial low-quality data. J Cloud Comput 13(1):62
25. Xu R, Li B, Li C, Joshi JBD, Ma S, Li J (2024) Tapfed: threshold secure aggregation for privacy-preserving federated learning. IEEE Trans Dependable Secur Comput 21(5):4309–4323
26. Jebreel NM, Domingo-Ferrer J, Blanco-Justicia A, Sánchez D (2022) Enhanced security and privacy via fragmented federated learning. IEEE Trans Neural Netw Learn Syst

Real-World Applications and Use Cases: Blockchain-Federated DeepFex in Media, Healthcare, and Finance

Mamta

Abstract Real-world applications are emerging that integrate blockchain, federated learning, and deepfakes to transform content authentication for media, healthcare, and other sectors. All of these technologies are combined to solve the challenges of deepfake detection while preserving privacy and security requirements. This chapter describes examples of real-world implementations of blockchain-federated deepfakes across three broad industries. The research methodology assesses three industry-specific applications by analyzing social media verification systems, hospitals' medical image security, financial institutions' authentication frameworks, as well as system architectures, privacy protection solutions, and performance measurements, forming a cost efficiency assessment in the study. In each case, measurable results, solutions, and integration challenges were examined in the research framework. The case study findings indicate successful integrated solutions across all three sectors. This system outperformed more traditional application model practices for media applications when it came to guaranteeing clean verification of content authenticity and reducing the number of false alerts for deepfake detection. Healthcare system implementations have strengthened medical imaging security and telemedicine protection while maintaining HIPAA regulatory compliance. Fintech results in better fraud detection applications and improved customer verification in the financial sector. In addition, common implementation patterns across industries were identified, and reusable solutions that can be leveraged in future deployments. This study demonstrates that deepfake solutions supported by blockchain federation provide viable ways to deploy these technologies in real-world scenarios. This framework ensures that organizations have functional support for expanding their project deployments and immutable features. The lessons learned from the performance numbers are a solid foundation for future use. These discoveries will help in the advancement of secure and privacy-preserving methods of deepfake detection in various fields.

Mamta (✉)
Department of CSE, Chandigarh University Mohali, Punjab, India
e-mail: mamta.airesearch@gmail.com

A. Kumar et al. (eds.), *Blockchain and Federated Learning Synergy for Privacy-Focused DeepFex Solutions*, Blockchain Technologies,
https://doi.org/10.1007/978-981-95-1394-9_8

Keywords Blockchain · Deepfake detection · Federated learning · Privacy preservation · Industry implementation · Security framework

1 Introduction

The movement of deepfake technology forward has unprecedented challenges for different spheres requiring sophisticated, dependable, and scalable identification [1]. This enables the integration of blockchain technology into federated learning (FL) for DeepFex (deepfake detection) solutions and thus solves these challenges. This chapter discusses the practical implementation of these integrated solutions in several media, healthcare, and financial sectors and concentrates on implementation and efficacy in real-world cases.

1.1 Real-World Implementation Challenges

(i) **Technical Challenges**

The implementation of blockchain-federated DeepFex solutions faces several technical hurdles. In practice, real-time analysis of high-rate streams of data involves large amounts of computation and efficient algorithms [2]. The integration of blockchain's distributed ledger with FL systems demands careful architectural considerations to maintain both performance and security. Storage optimization becomes crucial when dealing with large-scale multimedia content, especially in sectors handling sensitive information.

(ii) **Privacy Concerns**

Privacy preservation continues to be a major concern in all implementation sectors. It is also required for healthcare organizations to comply with HIPAA regulations while implementing DeepFex solutions [3]. For financial institutions to guard sensitive customer data and remain transparent, they need to. However, the rights of users of the media platforms need to be respected while verifying the authenticity of the content posted on the media platforms. Such privacy requirements mandate a complex distribution of encrypted data and protocols for handling encrypted data.

(iii) **Scalability Issues**

This is because as organizations grow and expand their operations, scalability becomes a very important factor. We must deal with the growing amount of data and not drop performance or security. However, its integration with cloud infrastructure introduces problems of integration while maintaining the advantages of the decentralized benefits.

(iv) **Integration with Legacy Systems**

Most organizations are running existing infrastructure and legacy systems. However, the integration of these modern blockchain-federated solutions within existing workflows is very difficult. However, the system should provide full compatibility with the aforementioned systems with uninterrupted data flow.

(v) **Cost Considerations**

Implementation costs can fall under hardware infrastructure, software development, maintenance, and training. Budget constraints make it necessary to balance between the need for advanced security features and the benefits they can bring to the business. While energy consumption for blockchain operations and spending on storage of data need to be kept in mind for long-term sustainability, the operating costs will be looked into for long-term support.

1.2 Objectives

- Examine sector-specific case studies
- Assess metrics of performance and key indicators of success
- Recognize new problems and promising approaches
- Use practical guidelines for implementation
- Evaluate future growth opportunities.

2 Literature Review

Blockchain technology has been examined by some researchers in the context of its adoption by the healthcare system and its potential to transform it. One such study provided a blockchain perspective to address the COVID-19 pandemic, including controlling the spread of the virus and mitigating systemic impacts [4]. A different review that involved 60 distinct studics among 1,647 reviewed ones described major health scenarios using blockchain. In addition, a systematic review included 85,375 articles, 415 of which were included, and conducted a meticulous review of the application of blockchain in healthcare [5]. Also, a few authors have been conducting modern papers on current apps and developments in the health industry, for example, solid approaches, current companies, issues, and prospects [6]. There is also evidence of a few studies on blockchain applications in clinical studies, with privacy, data integrity, and security being the most highlighted findings [7]. Another review analyzed the most recent developments of blockchain in healthcare, highlighting applications, challenges, and future perspectives [8]. Some studies have also specifically focused on the role of blockchain in EHR systems and its use for dealing with difficulties in the context of the coronavirus (COVID-19) pandemic as well [9]. A summary of blockchain concepts and healthcare applications identified existing gaps and challenges in the literature. The combination of blockchain and deep learning has also attracted attention. Some literature discusses the advantages of blockchain in the field of health, especially the security and precision of data [10]. Some corresponding classifications have been made in seven thematic taxonomies in some reviews, which highlight the merits of the combination of these technologies [11]. Another such creative method bolstered the Ayushman Bharat Digital Mission suggestion by integrating blockchain for increased transparency, auditability, and traceability [12]. The combination of blockchain and FL has been studied extensively, which allows privacy and security improvement in a range of applications. One paper takes a closer look at the FL integration with PETs and blockchain, especially concentrating on the technical aspects of blockchain in the context of FL frameworks [13]. Another work advocates a hybrid solution that integrates FL and blockchain to protect data privacy, integrity, and accountability, especially in IoT-based health care. Likewise, a systematic review on the integration of blockchain and FL for healthcare enumerates the pros, cons, and future opportunities. It's not just healthcare that is taking advantage of blockchain applications [14]. A study addresses use cases of blockchain in government, food supply, education, and healthcare, and reports implementation challenges [15]. Another systematic review specifically evaluates blockchain applications in the healthcare sector, focusing on challenges and opportunities within the healthcare network [16]. In addition, nonfinancial related blockchain applications have been covered in the media, serving as a case of using innovation and challenges to the growing pains faced by new technology. Use of blockchain in health care has become particularly known for improving security and access to patient data. A few papers addressed the scarcity of research at the confluence of blockchain and healthcare, but claim that blockchain will transform healthcare [17]. All such research shows that blockchain is considered an emerging technology for promising applications in solving healthcare data privacy issues.

3 Methodology

Figure 1 depicts the framework of implementation, showing stages of the process and key aspects of the implementation process. The flowchart operates in the manner of a protocol to ensure logical progression and consistency with the aim of the study [1].

3.1 Overview of Research Structure and Approach

The study framework will explore practical use cases of BFD in media, healthcare, finance, and so on. It takes a hybrid approach in which the qualitative nature of real-world deployment is combined with the quantitative performance evaluation [18].

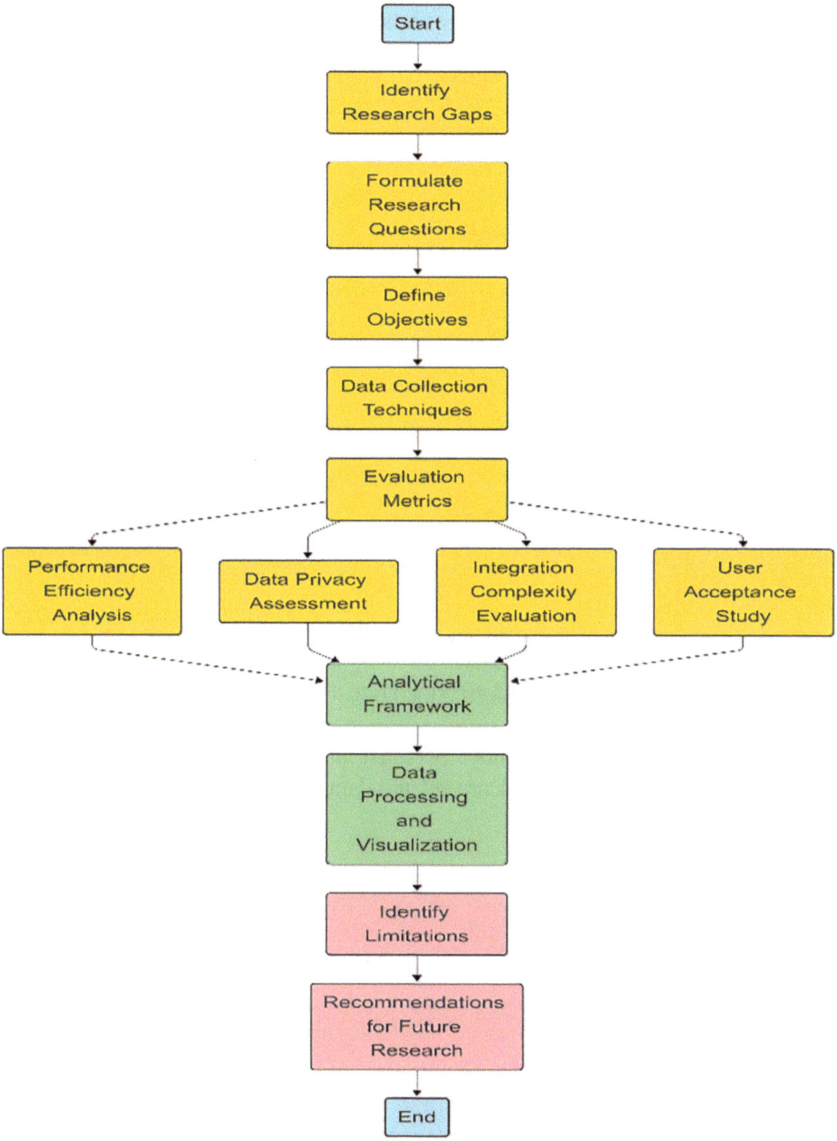

Fig. 1 Implementation framework for blockchain-federated DeepFex

The study seeks to narrow the gap between theory and practice by determining the main implementation issues and the best approaches.

The architecture consists of three main stages:

(i) **Pilot phase**: A comprehensive review of the literature to identify previous research and gaps.

(ii) **Collected data stage**: Collecting primary (academic papers, case studies) and secondary data (industry reports).
(iii) **Analytical stage**: The considerations of data processing, which is executed to check the efficiency and practicability of blockchain-federated DeepFex in various fields.

3.2 Methods Adopted for Data collection and Preprocessing

Data collection data is collected from reputable sources, in both quantitative and qualitative forms, to include academic libraries, such as Ebsco Academic Search and Science Direct, case study databases, technical reports, and industry publications. The collected dataset is classified into three primary types of applications, namely, media, health, and finance.

The corresponding procedure to ensure data quality and fitness-for-use is as follows:

(i) **Data cleaning**: Deleting inconsistencies, duplicates, and incomplete data to preserve data quality.
(ii) **Normalize data**: Uniform data types, units, and encoding systems to make the data consistent between datasets.
(iii) **Preprocessing**: Raw data transformed into a well-structured format applicable for blockchain and FL.
(iv) **Data identification**: To remove personal and sensitive data to adhere to data privacy laws, such as healthcare-related datasets.

3.3 Evaluation Metrics

The effectiveness of blockchain-federated DeepFex implementations is estimated using the following metrics:

(a) **Performance Efficiency**

- **Throughput**: Number of transactions or amount of data that can be processed in a unit of time.
- **Latency**: It measures the response time in scenarios such as media content verification and patient data access in healthcare applications.
- **Accuracy**: Measures the accuracy of processing data of applications (interfacing with data), which can be critical in real-time decision-making applications.

(b) **Data Privacy and Security**

- **Confidentiality**: It measures how the system secures sensitive data, specifically in healthcare scenarios.

- **Integrity**: Provides a mechanism to verify original data has not been corrupted or altered during transmission.
- **Resistance**: It measures the resistance against a cyber-attack or data leak, etc.

(c) **Scalability and Adaptability**

- **Resource utilization**: Evaluates the efficiency in computation and storage.
- **Scalability**: To test how the system performs under load and in different sectoral applications.
- **Cross-platform support**: Looks at how well the system can fit into an organization.

(d) **User Acceptance and Usability**

- **User of interface**: Surveys and feedback from users about the usability of the interface.
- **User satisfaction**: Measures perceived usefulness and ease of use of our blockchain-federated DeepFex solutions.
- **Compliance**: Verifies compliance with industry regulations like GDPR for health data.

3.4 Methods of Analysis and Visualization

The main analytical and visualization techniques in this research attempt to compare blockchain-federated DeepFex implementation between the media, health, and financial domains. The analysis starts with "descriptive analysis" to derive characteristics and to identify patterns of the data, and then with "comparative analysis" to compare performance indicators over sectors and reveal sector-specific bottlenecks (or lack of them). The performance and efficiency differences are statistically tested. Visualization methods are what-if scenario generation and demonstrations, including plot-based representations such as bar charts for performance comparison, pie charts for implementation challenges, line charts for efficiency over time, flowcharts for data processing flow, and network diagrams for interaction map of blockchain nodes [19]. A synthesis of the insights obtained with these visualizations shows associations and causes of the sectoral disparities in performance, and suggests optimisation strategies inferred from the visual analytical results.

4 Media Industry Applications

Fake news and sensationalism are rampant and threaten the industry. Solutions such as blockchain-federated DeepFex put in place robust means of authentication that span various media platforms, thus ensuring content authenticity and creator rights.

4.1 Content Authentication Systems

Deepfake content has limited business (media) implications, but to prevent further propagation, a content authentication system is the first line of defense. Combining blockchain immutable record-keeping with federated learning provides a privacy-preserving response to an emerging threat.

4.1.1 Social Media Content Verification

The massive amount of UGC in online social media calls for a real-time content validation mechanism. Our approach covers multiple levels, and deployment is based on DeepFex's detection algorithm that takes incoming content and new data, and at the same time maintains the user experience. As soon as the content gets uploaded, it is analyzed immediately, with a result that will be written on the blockchain for transparency and audibility [20]. It interlaces mechanisms for validating creator identity and tracking the origin of content, two pillars critical to ensuring the integrity of these platforms. A strong feedback mechanism allows the community to aid in verifying content while still maintaining systematic oversight. This approach allows for quick processing of content while still achieving high-accuracy deepfake detection.

4.1.2 News Media Integrity Verification

Stringent verification protocols are needed for news media organizations to maintain credibility and trust. Our Authentication Framework emphasizes deep provenance tracking and editorial attribution. Every piece of content has its unique digital signature that stamps the creation time, origin, and the changes made afterward. This system provides transparency in the editorial process while safeguarding content integrity. In metadata, authentication is vital for determining the authenticity of content [21]. It verifies creation time, where the data was created, and the equipment used for content creation, thus documenting a lifecycle of digital provenance for content. Editorial tracking keeps version control, as well as proper attribution (which is vital to journalistic standards and credibility).

4.1.3 Digital Content Ownership Tracking

Tracking digital content ownership is a powerful protection of intellectual property for creators and publishers. The ownership management system defines rights for content through registration, licensing, and recording information on the blockchain. By integrating content tracking, distribution, and monetization along with creator rights protection, this system empowers content creators, but it also allows efficient rights management.

Key Success Metrics

- Detection accuracy and verification reliability
- Performance of the system and response time
- Adoption rates and user experience

By creating unique identifiers, content fingerprinting allows this identification to be used across platforms and uses, ensuring accuracy. It uses automatic tracking based on smart contracts to track content distribution and ensures compliance throughout, such as verifying attributions while ensuring that content creators receive compensated royalties based on how their work has been used [22]. Some of the recent innovations that help determine the content usage pattern and track the revenue generation closely are Advanced Analytics, which provides insights to assist decision-making for the content strategy. The main actual issues with implementation are related to dealing with large amounts of data while being able to preserve the verification in a real-time manner. Solution architecture is designed leveraging distributed processing along with optimal storage techniques to meet these requirements. Performance optimization, on the other hand, is about maintaining fast response times but ensuring they are correct on verification. A complete method for authenticating content shows how blockchain-federated DeepFex solutions enhance the practicality of media applications. The system achieves vital status as a modern media operation tool because it maintains content integrity while fulfilling the needs of content creators and platform requirements.

4.2 Entertainment Industry Solutions

Digital technologies have brought major changes to the entertainment sector during this period.

These advancements create new security problems because cyber criminals attempt to harm intellectual property and digital assets and exploit well-known personalities. Businesses can find appropriate solutions by analyzing their specific challenges within this part [4]. Figure 2 shows a three-tier model that protects the entertainment industry. The system distributes protection measures into three distinct categories: movie/video protection, digital rights management, and deepfake protection. Every component in this approach contains specific technological methods that protect both digital content and celebrity identity information.

4.2.1 Movie and Video Content Protection

The primary objects for violation and unauthorized distribution in the entertainment industry are movies and video media. Significant technologies, along with protective strategies, need implementation to safeguard this content.

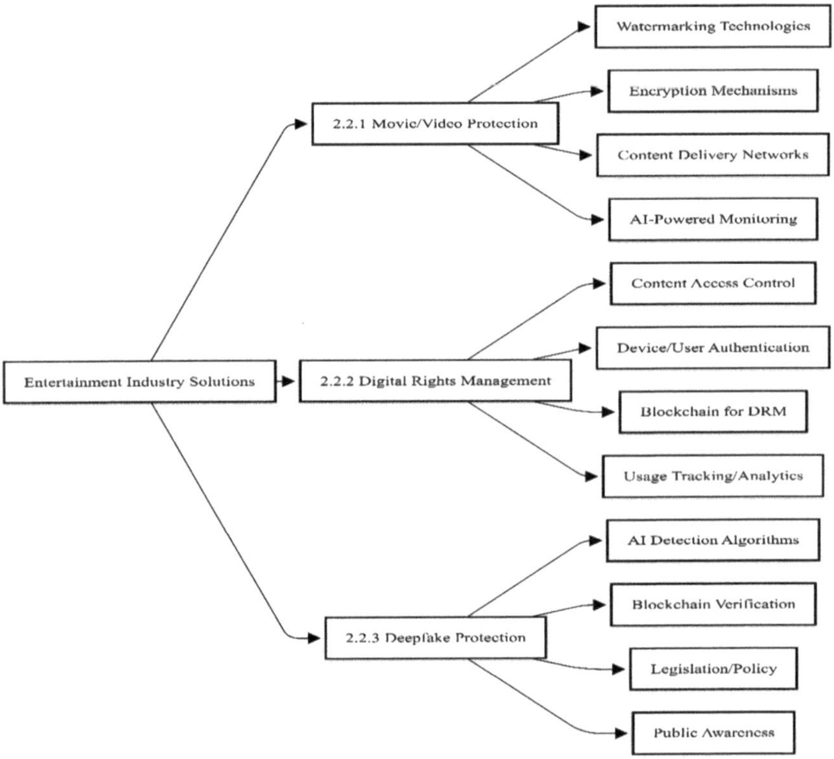

Fig. 2 Blockchain-federated DeepFex solutions for the entertainment industry: protecting media, rights, and authenticity

(i) **Watermarking Technologies**

- Watermarking technologies enable either hidden or visible authentication marks that allow tracking of original content.
- Forensic watermarking serves as a tool to track unauthorized content sharing and leaks.

(ii) **Encryption Mechanisms**

- Advanced Encryption Standard (AES) serves as a method to distribute protected content securely.
- The security system protects the encryption of content during both storage time and the transmission process.

(iii) **Content Delivery Networks (CDNs)**

- Secure streaming protocols, which include HTTPS and DRM-based CDNs, should be supported by the organization.

- Commercial content access depends on token-based authentication methods.

(iv) **AI-Powered Monitoring**:

- Artificial intelligence systems will operate as an internet scanning mechanism that detects unauthorized material and pirated websites.
- The system should automatically initiate takedown procedures on different platforms that host pirated content.

4.2.2 Digital Rights Management (DRM)

DRM serves as an essential framework that allows people to manage and secure digital content distribution [9]. Key approaches include:

(i) **Content Access Control**

- Subscription-based and pay-per-view systems should become implementation strategies for limiting content access.
- Users need to provide license keys for DRM systems to verify access permissions to particular content.

(ii) **Device and User Authentication**

- Multifactor authentication serves as an extra method to enhance security.
- Concepts protected by DRM remain linked to particular devices through a system that stops unauthorized distribution.

(iii) **Blockchain for DRM**

- The implementation of blockchain technology produces an unalterable database for managing ownership history and access privileges of content material.
- Smart contracts enabled through the platform serve for automated license execution as well as royalty payment management.

(iv) **Usage Tracking and Analytics**

- The system monitors content use to identify abnormal activities while stopping misuse of content.
- Distributors as well as producers gain detailed information about audience behavior through these systems.

4.2.3 Celebrity Deepfake Protection Mechanisms

Deepfake technology, while innovative, poses significant threats to the privacy and reputation of celebrities [23]. Protection mechanisms include:

(i) **AI Detection Algorithms**

- A combination of machine learning programs scans deepfake content through the assessment of minute video and audio inconsistencies.
- Commercial entities should establish a partnership with social media platforms to create automatic systems that detect deepfake content and trigger warning alerts.

(ii) **Blockchain Verification**

- The development of blockchain technology enables the establishment of digital signature verification systems to authenticate videos of celebrities.
- Context verification processes validate all material that deviates from the signature embedded within the data.

(iii) **Legislation and Policy Enforcement**

- Supporting the implementation of legal penalties that target deepfake makers who distribute unauthorized deepfakes.
- The platform needs to join regulatory bodies to establish standard operating procedures for content management services in user-generated content hosting.

(iv) **Public Awareness Campaigns**

- Public awareness campaigns must establish both the existence and harmful aspects of deepfake technology.
- Tools must exist for individuals to check if the content has authentic qualities or not.

4.3 Case Study: Social Media Platform Implementation

A social media platform adopts content protection measures as studied in this case. Deployment challenges, as well as architecture design and performance measurement, are components of the research [24].

4.3.1 Implementation Architecture

A solution architecture unites several elements to establish operations that combine security with efficiency.

(i) **Content Protection Module**

- A security system maintains the protection of uploaded videos by combining watermarking and encryption technologies.
- The system operates by running AI algorithms for both instant pirated content monitoring and automated removal procedures.

(ii) **DRM System**

- The system administrators control user permissions and stop users from giving away premium content without consent.
- Employs blockchain for transparent and secure license management.

(iii) **Deepfake Detection System**

- The application uses AI-based software to scan all uploaded media while identifying darkroom signs and indicating detected modifications.
- The network integrates verified third-party API solutions to perform additional verification tasks.

(iv) **User Experience Layer**

- Users experience smooth content usage through background security features integration. The system enables the prevention of unauthorized content sharing.
- Users have access to tools that enable them to report both suspicious content and content without authorized permission.

4.3.2 Performance Metrics

Multiple success indicators measure the performance of the solutions that were implemented.

(i) **Accuracy**

- The detection system displays two key attributes: first, it identifies deepfake content effectively with high precision while also identifying unauthorized uploads.
- The system reduces incorrect detections so that legitimate content fails to trigger the flagging system.

(ii) **Efficiency**

- The implemented solutions maintain platform performance at a consistent level, which allows users to have an uninterrupted experience.
- The system performs immediate checks on newly uploaded content to stop processing delays from occurring.

(iii) **User Retention**

- Enhanced user trust and satisfaction through robust content protection measures.
- The improved content protection system leads to higher subscription numbers because subscribers gain secure access to premium material.

(iv) **Scalability**

- The system displays high processing capacity for big content uploads, which delivers continuous performance benefits.
- Seamless integration with new technologies and updates.

4.3.3 Challenges and Solutions

Social media users face various hurdles when they stage content protection solutions [25]. The list of common problems, along with their solutions, consists of:

(i) **High Computational Overhead**

- **Challenge**: The current execution speed and encryption algorithm used in real-time operations require substantial computational power.
- **Solution**: A scalable cloud-based infrastructure deployment will solve the performance issues by optimizing algorithms.

(ii) **False Positives in Detection**

- **Challenge**: The system occasionally shows errors by incorrectly marking genuine content.
- **Solution**: The solution to improve algorithm accuracy consists of constant training using datasets from multiple sources.

(iii) **User Privacy Concerns**

- **Challenge**: The practice of checking user content to detect deepfakes creates privacy problems that users find objectionable.
- **Solution**: The protection of privacy and data compliance requires organizations to use privacy-protecting AI methods in addition to following relevant data protection laws.

(iv) **Adaptation to Evolving Threats**

- **Challenge**: The advancement of technology produces fresh methods for piracy and deepfake development.
- **Solution**: The organization should develop specific research and development groups that identify upcoming threats proactively.

(v) **Global Policy Variations**:

- **Challenge**: The digital content protection systems face different legal obstacles among regional frameworks in the United States.
- **Solution**: The organization will partner with local authorities to adjust implemented solutions that can fulfill different regional compliance regulations.

5 Healthcare Applications

Medical imaging combined with telemedicine operations is among the highest-priority security fields because it manages vital patient details that require protection. In this section, security aspects related to medical imaging, telemedicine, and healthcare network system implementation take center stage [26].

5.1 *Medical Imaging Security*

Medical imaging plays a vital role in diagnostics and treatment planning. Protecting the security of both diagnostic images and related healthcare information creates necessary conditions for trust and compliance in medical institutions.

5.1.1 Diagnostic Image Verification

Medical diagnostic images need absolute integrity protection and authentic verification to deliver correct patient medical assessments [27]. Key approaches include:

- Doctoral teams utilize image watermarking methods to add secret markers into medical pictures, helping detect modifications and confirming their origin.
- Organizations can guarantee diagnostic image integrity through blockchain technology while maintaining a tamper-evident record of such images.
- The system utilizes AI technology to discover unusual features in diagnostic pictures that signal potential tampering or corruption.

5.1.2 Patient Data Privacy

Patient healthcare data linked to medical imaging needs full protection because privacy standards require it, and patient trust depends on it. Strategies include:

- **Encryption**: Each patient data set must be encrypted for secure storage as well as transmission to prevent unauthorized access during the entire process.
- **Access control**: Role-based access control should be implemented to guarantee that authorized workers alone have access to sensitive patient information.
- **Compliance with HIPAA and GDPR**: The organization maintains complete adherence to HIPAA and GDPR requirements for protecting patient data, which prevents legal consequences.

5.1.3 Cross-Hospital Collaboration Frameworks

The integration of hospitals for patient care improvement demands safe data-sharing systems to safeguard mutual information platforms. Solutions include:

- **Interoperable systems**: A standards-based platform must be developed since it establishes security protocols for reliable hospital-to-hospital data transmission.
- **Federated learning**: The utilization of federated machine learning models enables healthcare facilities to exchange medical insights as FL protocols protect actual patient information from exposure.
- **Secure data sharing protocols**: Healthcare organizations should create protocols based on secure multi-party computation to protect data that needs to be shared.

5.2 Telemedicine Protection

Virtual healthcare through telemedicine has transformed patient care while creating complex difficulties regarding the protection of virtual medical sessions and medical records [28].

5.2.1 Remote Consultation Security

For preserving sensitive discussions and diagnoses, it is vital to protect remote consultation security. Measures include:

- **End-to-end encryption**: The modern age requires end-to-end encrypted channels to avoid eavesdropping during virtual medical interaction sessions.
- **Secure videoconferencing platforms**: The secure use of videoconferencing platforms exists for telemedicine because these platforms implement strong security features made specifically for remote medical encounters.
- **Session monitoring**: Healthcare professionals use session monitoring systems to protect patients from unauthorized entry barriers during consultation periods.

5.2.2 Patient Identity Verification

Patient identity verification needs to be exact to stop cases of both fraud and misdiagnosis. Techniques include:

- **Biometric authentication**: The healthcare industry can authenticate patients using various methods that include facial recognition in combination with fingerprint scans and voice authentication.
- **Two-factor authentication**: Security protection through two-factor authentication will verify access for authorized people who need to use telemedicine services.

5.2.3 Medical Record Authentication

Secure access to verified medical records enables treatment professionals to receive correct information for medical needs [29]. Key methods include:

- **Digital signatures**: A digital signature system serves to verify medical records because it confirms their authenticity.
- **Blockchain for record integrity**: Technology allows healthcare providers to use blockchain to maintain tamper-proof medical record histories to preserve their integrity.
- **Access logs**: The system should maintain thorough logs of medical record access events for better accountability measures.

5.3 Case Study: Hospital Network Implementation

The study investigates hospital network implementation of secure healthcare solutions by detailing their system structures alongside privacy methods and healthcare regulation adherence [30].

5.3.1 System Architecture

A hospital network system requires certain architectural elements, which include:

- **Centralized data repositories**: Making use of centralized systems secured with encryption and access controls to store patient records and medical images.
- **Distributed access points**: The healthcare system implements distributed access points through which medical staff members maintain secure access to information across different geographical positions while maintaining security integrity.
- **AI-driven security modules**: The implementation of artificial intelligence detects system activities while searching for unusual behaviors and stops unauthorized entry during actual-time operations.

5.3.2 Privacy Preservation Mechanisms

A system for guaranteeing patient privacy demands the implementation of technological solutions and procedural protocols.

- **Data anonymization**: The practice of data anonymization involves the removal of all person-identifying information that doctors call PII from healthcare records before statistical analysis or patient record sharing [31].
- **Consent management systems**: The system enables patient consent through which people gain authority to decide who sees their data, along with their authorized use purposes.

- **Secure cloud platforms**: Healthcare operators should use compliant healthcare-specific cloud platforms to store their data.

5.3.3 Compliance with Healthcare Regulations

Organizations must follow regulatory obligations because this practice protects patient data and prevents legal consequences.

- **HIPAA compliance**: Companies are required by law to follow HIPAA compliance for the protection of health records as per Health Insurance Portability and Accountability Act.
- **GDPR compliance**: Ensuring compliance with the General Data Protection Regulation for patients within the European Union.
- **Periodic audits**: The practice conducts routine audits, which help to discover security weaknesses and maintain adherence to developing regulatory standards.

6 Financial Sector Applications

Resistance to security breaches and fraud, together with efficiency needs in the financial sector, requires implementing advanced technological solutions. This section delves into banking security, insurance claim processing, and the implementation of these technologies in financial institutions.

6.1 Banking Security Solutions

Banks serve as leading targets during cyberattacks, along with fraudulent activities [32]. The risks are minimized by banking institutions through the implementation of robust security protocols.

6.1.1 Customer Verification Systems

Protecting the true identity of customers stands as an essential requirement for banking security systems to operate safely [33]. Key technologies include:

- **Biometric Authentication**: Customer verification now involves the use of three biometric systems that combine fingerprint analysis with facial and iris recognition proof.
- **Know your customer (KYC) compliance**: A thorough KYC implementation system provides customers with authentication while stopping possible identity fraud attempts.

- **Multifactor authentication (MFA)**: MFA systems require customers to activate their accounts with passwords followed by OTPs and verification through biometric methods.

6.1.2 Transaction Authentication

Authenticating transactions is made secure through established authentication systems that approve legitimate commands only [34]. Methods include:

- **Token-based authentication**: Transactions get confirmed by physical or digital tokens under token-based authentication.
- **Real-time OTP verification**: Sending one-time passwords to customers' registered devices for transaction approval.
- **Behavioral biometrics**: Behavioral biometrics utilizes user activity, such as typing habits and movement through the system, to track irregularities that occur throughout transactions.

6.1.3 Fraud Prevention Mechanisms

Proactive action must be taken to prevent fraudulent activities through several methods, including:

- **AI-powered fraud detection**: Machine learning model deployment to capture deviations from normal and flag suspicious activity
- **Geolocation monitoring**: Transactions need to be geographically monitored to identify irregularities within the location data.
- **Transaction limits**: The establishment of transaction limits functions to lower the impact that unauthorized activities could cause.

6.2 Insurance Claim Processing

Insurance organizations encounter difficulties when they need to validate claims, together with risk evaluations [35]. Automated and secured claim processing can facilitate faster operations and minimize fraud.

6.2.1 Document Verification Systems

Document verification processes require high efficiency to stop fraudulent claim activities. Technologies include:

- **OCR**: The service employs OCR to document feed in during verification.

- **Blockchain for document integrity**: The use of blockchain technology creates tamper-proof documented records of submitted documents through a system called blockchain for document integrity.
- **AI-driven verification**: Investigating documents for discrepancies or indications of forgery.

6.2.2 Automated Claim Authentication

The automation of claim authentication processes accelerates the claim handling speed while decreasing the likelihood of human errors. Approaches include:

- **Rule-based systems**: Automated claim assessment depends on rule-based systems that implement predefined assessment rules for determining eligibility.
- **Natural language processing (NLP)**: Analyzing claim descriptions and associated documents for verification.
- **Integration with external databases**: Claim authenticity verification takes place through cross-comparisons with external database sources such as medical records and vehicle registries.

6.2.3 Risk Assessment Frameworks

Data provides tools to the insurance industry for accurate risk appraisal, hence reducing losses and maximizing premiums [36]. Solutions include:

- **Predictive analytics**: Utilizing historical data to forecast the likelihood of claims being filed, along with related risks.
- **IoT data integration**: Making use of IoT devices, like telematics in cars or wearable health monitors, to determine risk levels.
- **Dynamic risk scoring**: The risk score is updated based on new data or events.

6.3 Case Study: Financial Institution Implementation

The following case study outlines the integration, performance, and security implementation of security and automation solutions within a financial institution [37].

6.3.1 Integration with Existing Systems

Seamless integration of new solutions with legacy systems can be tricky yet mandatory for business operations. Key steps include:

- **API-based integration**: API-based integration allows banks and insurance providers to link new applications with their conventional operating platforms through API interfaces.
- **Data migration strategies**: Ensuring secure and accurate data migration from old systems to new platforms.
- **Interoperability standards**: Adopting standards like ISO 20022 for smooth integration across financial systems.

6.3.2 Performance Evaluation

Solution performance evaluation provides the means to verify their operational success. Metrics include:

- **System uptime**: Monitoring the availability and reliability of integrated systems.
- **Transaction speed**: Transaction speed provides results that track the duration required for verifying and handling payments.
- **Fraud detection rates**: Assessing the accuracy and efficiency of fraud detection mechanisms.

6.3.3 Security Measures

Security measures that offer robust defense play a central role in safeguarding both financial data and operational procedures. These include:

- **End-to-end encryption**: The protection techniques used by the company secure data at every stage of its journey from creation to termination.
- **Regular penetration testing**: The detection of security system vulnerabilities through regular penetration tests leads to vulnerability mitigation.
- **Compliance with regulations**: The organization follows standards PCI DSS, GDPR, and local banking regulations as part of its legal compliance initiative.

7 Results and Discussions

This section presents the analysis and findings related to the implementation of blockchain-federated DeepFex across three key sectors: media, healthcare, and finance [38]. The analysis highlights sector-specific challenges, performance metrics, and insights into the practical integration of blockchain and FL techniques.

Key Findings

(i) **Media sector**: Demonstrates improved content verification but faces challenges related to integration and data processing speed.

(ii) **Healthcare sector**: Exhibits enhanced data privacy and patient data management, yet encounters issues with interoperability between healthcare systems.
(iii) **Finance sector**: Achieves higher transaction speed and fraud prevention efficiency, but maintaining data consistency remains challenging.

Performance Metrics Table

Table 1 describes the performance measures of blockchain-federated DeepFex in terms of media, healthcare, and finance. Content validation, data privacy, transaction/channel speed, integration effort, and data consistency are some of the critical parameters. Healthcare is best in the privacy of data, and finance ranks highest in transaction speed. Media achieves effective CVC but suffers from integration complexity.

Figure 3 shows the performance comparison of the blockchain-federated DeepFex targeted for three central domains: media, health, and finance. Its emphasis is on three critical areas: content verification, data privacy, and transaction velocity. Each has different aspects of excellence, such as healthcare data privacy and financial transaction speed. Media, by contrast, has strong verification features. The stacked bar graph clearly shows the distribution and comparative performance of each metric in the sectors, reinforcing the importance of case-specific optimization for different industrial sectors.

Analysis

- The media sector exhibits high content verification, though the integration process is complex and may affect performance.
- In the healthcare domain, while data privacy metrics are significantly high, interoperability across various healthcare systems presents a notable challenge.
- The finance sector shows faster transaction speeds compared to other sectors, but ensuring consistent data accuracy remains a primary focus.

Interpretation

The comparative analysis suggests that the healthcare sector has the most robust data privacy framework, while the finance sector leads in transaction speed. The media sector, despite its strong content verification metrics, faces integration challenges

Table 1 Performance metrics comparison of blockchain-federated DeepFex across media, healthcare, and finance sectors

Sector	Content verification (%)	Data privacy (%)	Transaction speed (%)	Integration complexity	Data consistency
Media	85	80	70	High	Medium
Healthcare	88	92	75	Medium	High
Finance	80	85	88	Medium	Low

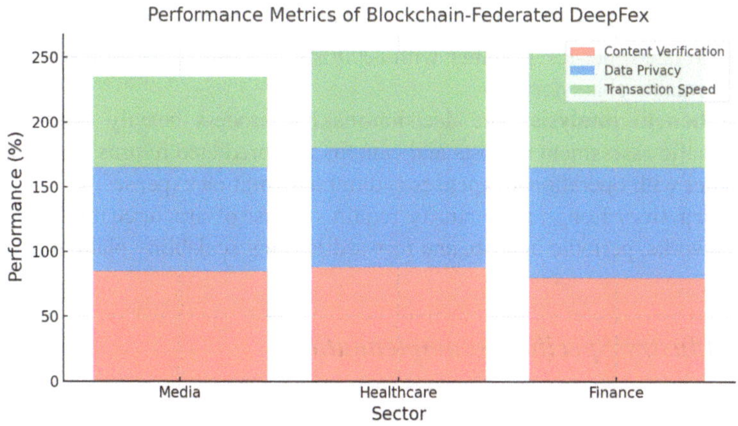

Fig. 3 Performance analysis of blockchain-federated DeepFex in various domains, including media, healthcare, and finance

that require streamlined approaches. To enhance efficiency, targeted improvements should be implemented, considering each sector's unique demands.

Discussion

The results demonstrate that Blockchain-Federated DeepFex can effectively address domain-specific challenges in media, healthcare, and finance. Nevertheless, scalability and integration remain crucial for broader adoption. Further exploration is needed to optimize the framework for diverse use cases and to develop standardized integration techniques that reduce complexity.

8 Conclusion

This chapter ends with a complete synthesis of learned knowledge and specific industry suggestions in addition to future study guidelines.

9 Summary of Key Findings

This analysis revealed numerous essential findings throughout the study. The systematic research emphasizes that organizations require an organized structure when implementing technology systems across different fields.

- **Cross-industry patterns**: Various sectors display parallel, exceptionally designed characteristics because they experience regular operational obstacles, including data privacy problems and problems with legacy systems, and cyber threats.

- **Performance metrics**: Efficiency assessments of systems depend heavily on speed-related metrics, together with accuracy levels and resource usage, along with scalable characteristics.
- **Cost–benefit analysis**: The decision-making process heavily depends on a systematic assessment of costs and benefits and predicted returns of investment, together with operational expenses and implementation expenses.
- **Best practices**: Long-term maturity requires the use of structured implementation frameworks, periodic upkeep, and forward-looking scalability planning.

9.1 Industry-Specific Recommendations

We have tailored recommendations for each industry, including:

- **Healthcare**: Emphasis on strong patient data privacy mechanisms and compliance with healthcare regulations will be necessary to build trust and ensure legal adherence.
- **Financial sector**: Fraud prevention and transaction authentication should have the utmost importance.
- **Entertainment**: Upgrade digital rights management and implement AI-powered deepfake detection systems to safeguard intellectual property and reputation.
- **Cross-industry**: Utilizing reusable solutions involving standard encryption protocols and scalable AI models to reduce expenses and enhance productivity

9.2 Future Research Directions

Additional research is needed to develop original findings within these areas.

- **Emerging technologies**: Diving into new technologies, quantum computing, blockchain, and advanced AI to see how they can solve industry-specific problems.
- **Interdisciplinary approaches**: Bringing together knowledge from a range of fields to create new ways to solve problems.
- **Longitudinal studies**: Scientists should pursue long-term research that enables analysis of system effectiveness throughout time.
- **User-centric designs**: The studies focus on two priority domains to create design outcomes advantageous for all users, regardless of their background.

References

1. Al-Nbhany W, Zahary A, Al-Shargabi A (2024) Blockchain-IoT healthcare applications and trends: a review. IEEE Access 12:4178–4212. https://doi.org/10.1109/ACCESS.2023.3349187

2. Villarreal E, García-Alonso J, Moguel E, Alegría J (2024) Blockchain for healthcare management systems: a survey on interoperability and security. IEEE Access 11:5629–5652. https://doi.org/10.1109/ACCESS.2023.3236505
3. Deepak,Gulia P, Gill N, Yahya M, Gupta P, Shukla P, Shukla P (2024) Exploring the potential of blockchain technology in an IoT-enabled environment: a review. IEEE Access 12:31197–31227. https://doi.org/10.1109/ACCESS.2024.3366656
4. Chaddad A, Wu Y, Desrosiers C (2024) Federated learning for healthcare applications. IEEE Internet Things J 11:7339–7358. https://doi.org/10.1109/JIOT.2023.3325822
5. Alsamhi S, Myrzashova R, Hawbani A, Kumar S, Srivastava S, Zhao L, Wei X, Guizan M, Curry E (n.d) Federated learning meets blockchain in decentralized data-sharing: healthcare use case. IEEE Internet Things J. https://doi.org/10.1109/jiot.2024.3367249
6. Pasupuleti MK (2024) Decentralized innovation: transforming finance, real estate, healthcare, and energy with blockchain, pp 12–25. https://doi.org/10.62311/nesx/46011
7. Myrzashova R, Alsamhi SH, Hawbani A, Curry E, Guizani M, Xi W (2024) Safeguarding patient data-sharing: blockchain-enabled federated learning in medical diagnostics. IEEE Trans Sustain Comput 1–15. https://doi.org/10.1109/tsusc.2024.3409329
8. Aziz K, Dua S, Gupta P (2024) An explainable and comprehensive federated deep learning in practical applications, pp 109–130. https://doi.org/10.1201/9781032694870-9
9. Baliker C, Baza M, Alourani A, Alshehri A, Alshahrani H, Choo K (2024) On the applications of blockchain in FinTech: advancements and opportunities. IEEE Trans Eng Manage 71:6338–6355. https://doi.org/10.1109/TEM.2022.3231057
10. He Q, Feng Z, Fang H, Wang X, Zhao L, Yao Y, Yu K (2024) A Blockchain-based scheme for secure data offloading in healthcare with deep reinforcement learning. IEEE/ACM Trans Netw 32:65–80. https://doi.org/10.1109/TNET.2023.3274631
11. Younis H, Bwaliez O, Al-Okaily M, Tanveer M (2024) Revolutionizing supply chain management: a critical meta-analysis of empowerment and constraint factors in blockchain technology adoption. Bus Process Manag J 30:1472–1500. https://doi.org/10.1108/bpmj-10-2023-0805
12. Salim S, Turnbull B, Moustafa N (2024) A blockchain-enabled explainable federated learning for securing internet-of-things-based social media 3.0 networks. IEEE Trans Comput Soc Syst 11:4681–4697. https://doi.org/10.1109/TCSS.2021.3134463
13. Udeh E, Amajuoyi P, Adeusi K, Scott A (2024) The role of blockchain technology in enhancing transparency and trust in green finance markets. Financ Account Res J. https://doi.org/10.51594/farj.v6i6.1181
14. Myrzashova R, Alsamhi S, Shvetsov A, Hawbani A, Wei X (2023) Blockchain meets federated learning in healthcare: a systematic review with challenges and opportunities. IEEE Internet Things J 10:14418–14437. https://doi.org/10.1109/JIOT.2023.3263598
15. Stephanie V, Khalil I, Atiquzzaman M, Yi X (2023) Trustworthy privacy-preserving hierarchical ensemble and federated learning in healthcare 4.0 With blockchain. IEEE Trans Industr Inf 19:7936–7945. https://doi.org/10.1109/TII.2022.3214998
16. Liu Y, Yu W, Ai Z, Xu G, Zhao L, Tian Z (2023) A blockchain-empowered federated learning in healthcare-based cyber-physical systems. IEEE Trans Netw Sci Eng 10:2685–2696. https://doi.org/10.1109/TNSE.2022.3168025
17. Ramzan S, Aqdus A, Ravi V, Koundal D, Amin R, Ghamdi M (2023) Healthcare applications using blockchain technology: motivations and challenges. IEEE Trans Eng Manage 70:2874–2890. https://doi.org/10.1109/TEM.2022.3189734
18. Alzubi J, Alzubi O, Singh A, Ramachandran M (2023) Cloud-IIoT-based electronic health record privacy-preserving by CNN and blockchain-enabled federated learning. IEEE Trans Industr Inf 19:1080–1087. https://doi.org/10.1109/TII.2022.3189170
19. Khot MS, Madalgi JB (2023) Application of blockchain in healthcare: overview. Int J Sci Technol Eng 11(8):280–286. https://doi.org/10.22214/ijraset.2023.55186
20. Afaq YM, Manocha A (2023) Blockchain and deep learning integration for various applications: a review. J Comput Inf Syst 1–14. https://doi.org/10.1080/08874417.2023.2173330
21. Mukherjee A, Halder R, Chandra J, Shrivastava S (2023) HealthChain: a blockchain-aided federated healthcare management system. Int Conf Blockchain 1–5. https://doi.org/10.1109/ICBC56567.2023.10174916

22. Sharma P (2023) Blockchain: the digital ledger fighting fraud in healthcare. J Med Healthc 1–2. https://doi.org/10.47363/jmhc/2023(5)e110
23. Krichen M, Ammi M, Mihoub A, Almutiq M (2022) Blockchain for modern applications: a survey. Sensors (Basel, Switzerland) 22. https://doi.org/10.3390/s22145274
24. Sharma P, Namasudra S, Chilamkurti N, Kim B, Crespo G (2022) Blockchain-based privacy preservation for IoT-enabled healthcare system. ACM Trans Sens Netw 19:1–17. https://doi.org/10.1145/3577926
25. Houda Z, Hafid A, Khoukhi L, Brik B (2023) When collaborative federated learning meets blockchain to preserve privacy in healthcare. IEEE Trans Netw Sci Eng 10:2455–2465. https://doi.org/10.1109/TNSE.2022.3211192
26. Lakhan A, Mohammed M, Nedoma J, Martínek R, Tiwari P, Vidyarthi A, Alkhayyat A, Wang W (2022) Federated-learning-based privacy preservation and fraud-enabled blockchain IoMT system for healthcare. IEEE J Biomed Health Inform 27:664–672. https://doi.org/10.1109/JBHI.2022.3165945
27. Jabbar R, Dhib E, Said A, Krichen M, Fetais N, Zaidan E, Barkaoui K (2022) Blockchain technology for intelligent transportation systems: a systematic literature review. IEEE Access 10:20995–21031. https://doi.org/10.1109/ACCESS.2022.3149958
28. Odeh A, Keshta I, Al-Haija Q (2022) Analysis of blockchain in the healthcare sector: application and issues. Symmetry 14:1760. https://doi.org/10.3390/sym14091760
29. Habib G, Sharma S, Ibrahim S, Ahmed I, Qureshi S, Ishfaq M (2022) Blockchain technology: benefits, challenges, applications, and integration of blockchain technology with cloud computing. Futur Internet 14:341. https://doi.org/10.3390/fi14110341
30. Ray P, Dash D, Salah K, Kumar N (2021) Blockchain for IoT-based healthcare: background, consensus, platforms, and use cases. IEEE Syst J 15:85–94. https://doi.org/10.1109/JSYST.2020.2963840
31. Hussien H, Yasin S, Udzir N, Ninggal M, Salman S (2021) Blockchain technology in the healthcare industry: trends and opportunities. J Ind Inf Integr 22:100217. https://doi.org/10.1016/J.JII.2021.100217
32. Tandon A, Kaur P, Mäntymäki M, Dhir A (2021) Blockchain applications in management: a bibliometric analysis and literature review. Technol Forecast Soc Chang 166:120649. https://doi.org/10.1016/J.TECHFORE.2021.120649
33. Miglani A, Kumar N (2021) Blockchain management and machine learning adaptation for IoT environment in 5G and beyond networks: a systematic review. Comput Commun 178:37–63. https://doi.org/10.1016/J.COMCOM.2021.07.009
34. Azbeg K, Ouchetto O, Andaloussi S, Fetjah L (2021) A taxonomic review of the use of IoT and blockchain in healthcare applications. Irbm. https://doi.org/10.1016/J.IRBM.2021.05.003
35. Xie Y, Zhang J, Wang H, Liu P, Liu S, Huo T, Duan Y-Y, Dong Z, Lu L, Ye Z (2021) Applications of blockchain in the medical field: narrative review. J Med Internet Res 23(10). https://doi.org/10.2196/28613
36. Barai AK, Bhadoria RS, Bagwari J, Perl I (2021) A blockchain-based federated learning: concepts and applications. IGI Global, pp 158–177. https://doi.org/10.4018/978-1-7998-5876-8.CH008
37. Fekih R, Lahami M (2020) Application of blockchain technology in healthcare: a comprehensive study. Impact Digit Technol Public Health Dev Dev Countries 12157:268–276. https://doi.org/10.1007/978-3-030-51517-1_23
38. Shi S, He D, Li L, Kumar N, Khan M, Choo K (2020) Applications of blockchain in ensuring the security and privacy of electronic health record systems: a survey. Comput Secur 97:101966–101966. https://doi.org/10.1016/j.cose.2020.101966

Federated Learning for Enhanced Security in 6G-Enabled Internet of Medical Things Systems

M. Martina Jose Mary, R. Shyamala Devi, M. Devaraj, and Matthew Olusegun Adigun

Abstract The Internet of Medical Things (IoMT) represents a convergence of sensors, healthcare devices, and the Internet of Things (IoT), aiming to enhance healthcare services intelligently. Despite its potential, security and privacy concerns have impeded widespread integration, leading to a scarcity of high-quality IoMT datasets. Federated learning (FL), an emerging distributed learning method, offers enhanced security and privacy, making it beneficial for IoMT networks and smart healthcare systems (SHS). In the context of safeguarding critical networks against evolving cyber threats, intrusion detection systems (IDS) and ransomware have become indispensable. However, existing security models pose challenges and are often computationally expensive, particularly for resource-limited medical IoT devices. In this paper, we propose a privacy-preserving FL-based IDS model, designed to identify cyberattacks within IoMT networks. This model enhances the data privacy and security in IoMT to identify the cyberattacks in wearable devices, implantable devices, medical equipment, and logistic systems. The study demonstrates that integrating FL with 6G capabilities results in a robust and expandable security framework for medical infrastructures, protecting patient data while preserving functionality. All things considered, this strategy provides an intelligent, safe, and well-prepared solution for the healthcare systems of the future.

Keywords Internet of Medical Things (IoMT) · Federated learning (FL) · Data privacy

---

M. Martina Jose Mary (✉) · R. Shyamala Devi · M. Devaraj
Department of Computer Science and Engineering, IFET College of Engineering, Villupuram, India
e-mail: maryjosejerlin@gmail.com

M. O. Adigun
Department of Computer Science and Engineering, University of Zululand, Kwadlangezwa, South Africa
e-mail: adigunm@unizulu.ac.za

A. Kumar et al. (eds.), *Blockchain and Federated Learning Synergy for Privacy-Focused DeepFex Solutions*, Blockchain Technologies,
https://doi.org/10.1007/978-981-95-1394-9_9

1 Introduction

The new paradigm called the Internet of Things, or IoT, is built around the idea of the connection of physical objects with sensors in the digital world. A robust, fully connected intelligence network is required due to the growth of connected devices brought about by the introduction of 6G technology [1, 2]. The current transformation of the healthcare system from Industrial Revolution 4.0, which enhances healthcare delivery and reliability [3]. Wireless communication, along with wireless sensors, wireless broadband, portable gadgets, and cloud computing, are the primary factors that led to these changes. IoMT functions as a network that consists of medical devices, healthcare systems, and wearables alongside sensors and medical supplies to contribute to various healthcare applications, including Tness Pro, along with remote wellness initiatives, chronic care programs, and cost-efficient programs for elderly care that result in fast responses and high-quality medical treatment [4, 5]. The Internet of Medical Things (IoMT) functions as a network of clinical systems, including medical equipment and wearable technology with sensors, to support diverse healthcare applications such as Tness Pro, together with remote health monitoring programs, long-term illnesses, affordable senior care, prompt response measures, and superior treatment quality. Any device connected to a network enables us to extract data, and any networked device allows the same data collection abilities [6]. Applications from the IoMT field enable connection to networking systems and therefore allow data acquisition from any source. In recent years, numerous security breaches have been carried out by external users who work with the data. IoT devices have been the target of several attacks (Table 1). Attackers frequently target medical devices and organizations. Machine learning experts are currently interested in transfer learning and federated learning as two approaches to deep learning. Transfer learning serves as an approach that utilizes already trained models to resolve new issues with reduced amounts of input data [7].

Federated learning serves as an excellent approach for medical cyber-physical systems because it enables algorithm training across distributed platforms that store local data without any information exchange [8]. The main contributions of this paper are a distinctive federated model and an approach that enables the creation of a tailored global model utilizing local models by preserving model privacy, along with improved training efficiency. Employing the proposed federated learning-based intrusion detection systems (IDS) enables the protection of the patient healthcare system devices [4]. The proposed model demonstrates equivalent results to distributed learning systems while performing on the CICIDS2017 dataset through simulation. Healthcare networks will gain improved security performance through the implementation of federated learning in IoMT. Secure patient data storage occurs due to edge computing and protected model aggregation, which decreases security breaches together with unauthorized access [6]. Federated learning delivers protected individual healthcare solutions to patients, together with IoT healthcare applications. The proposed scalable model allows its application to various healthcare environments, from remote patient monitoring to smart hospitals. The current evolution of

Table 1 The timeline for cyber incident

Timeline	Cyber incident
August 13, 24	After the 3 AM ransomware operation stole and released patient data, Kootenai Health reported an additional data breach that involved more than 464,000 patients. The 22 GB collection of compromised details is also out there in the public domain, which means that any other cyber attackers can obtain the stolen info and implement it into different attacks
August 4, 24	INC ransomware-McLaren Health Care in Michigan confirmed that the outages affecting computer and phone systems were due to a cyberattack the organization had suffered midweek
June 03, 2024	The ransomware attack led several large hospitals in the National Health Service of London to shut down most of their health care services provided by Synnovis, a company that offers pathology and diagnostic services. The chances of people with the same blood type encountering in the hospitals will also be lowered. The postponement of operations like kidney transplants and chemotherapy for cancer patients was inevitable
May 14, 2024	The Singing River Health System is warning the public that they now assume a total of 895,204 people received treatment at the health center after the ransomware attack in August 2023. The threat actors are alleged to have released 80% of the data from the Singing River breach, as they claimed and displayed a catalogue of 420,766 files with a total size of 754 GB
May 06, 2024	According to MedStar Health, when an outsider hacked the three employees' emails and files, about 184,000 individuals' information was probably exposed. Emails and files also contained names, mailing addresses, dates of birth, dates of service, names of provider, and their health insurance
April 17, 2024	A reported ransomware attack in December 2023 affected almost 185,000 individuals from a nonprofit Michigan healthcare organization Cherry Street Services According to a filing to a regulatory body, the attack happened on the 21st of the month, and they only noticed the intrusion on the 24th, the holiday the event occurred on the villains had accessed credit card details, among other financial data like security codes or passwords
April 02, 2024	In a notice, the healthcare organization published on their website, the hack exposed the details of over 820,000 patients
March 28, 2024	In the spring of last year, Harvard Pilgrim Health Care said a ransomware attack, and today it has issued a statement saying that it has had more patients than it thought were affected by this incident. A ransomware group yet to be named has targeted the New England health insurance company since April 17, 2023, rendering service for days. Since then, the company has provided several breach notification letters to Maine regulators, with the last letter raising the number to 2,860,795, or about 12% higher
February 11, 2024	Of the 100 hospitals, 25 hospitals admitted that their data had been encrypted by the attackers; 75 other facilities, which also use HIS, took their systems offline to contain the problem while investigations are continuing. The Romanian Ministry of Health said that the attackers demanded 3.5 BTC (approximately €157,000 in ransom)
January 08, 2024	Failing to leave enigmatic status quo messages, the LockBit ransomware operation has posted blackmailing notes, stating about the attack on the Capital Health hospital network in November 2023 and threatening to publish the stolen data and chats with the company

cyber threats makes federated learning an effective method to enhance IoMT security frameworks because it leads to stronger digital healthcare ecosystems [9].

2 Privacy and Security

One of the challenges facing the healthcare sector is the safe and effective monitoring of patients' critical data [10]. The quality of services provided and patient safety have improved in the healthcare sector as a result of the adoption and use of the IoT [10]. However, the sensitive nature of patient clinical data and the difficulty of sharing it across heterogeneous networks and IoT devices make IoT use in healthcare difficult [7]. A primary aspect of federated learning is the technical issue of data privacy and security. These key areas are critical in 6G-enabled IIoMT systems because 6G-IIoMT involves concentrated patient data and vital healthcare apparatuses. FL is essential for solving these issues because it allows implementing the concept of local ML, where IIoMT devices work with data and share only deltas to the learned model with a common collector; this way, the risk of data leakage and unauthorized access is minimized [11]. This technique is General Data Protection Regulation (GDPR) and Health Insurance Portability and Accountability Act (HIPAA) compliant while ensuring that patients' data is not exposed during the process of transfer or even centralized storage. From the security aspect, FL enhances intrusion detection capabilities by training models from devices separated across a network, enabling early identification of prospective threats such as DDoS, ransomware, and other zero-day threats [12]. They are numerous cyberattacks have occurred in internet-connected medical devices, including wearable devices, implantable devices, medical equipment, and logistic systems. Some of the cyberattacks that have occurred in internet medical devices include data interception, third-party accessibility, data leakage, device hijacking, DDoS, data breaches, operational disruptions, ransomware, man-in-the-middle attacks, and botnets [13]. The healthcare industry finds federated learning applicable because it provides strong model performance alongside rigid privacy compliance. Healthcare facilities can develop intuitive systems by restricting their patient information to local storage areas, which prevents disclosure of confidential records. The decentralized data storage model diminishes security risks, so data breaches result in fatal consequences [13]. The security profile of IDS receives benefits through the approach of federated learning. Early threat detection of ransomware as well as DDoS attacks and future cyber threats becomes possible through modeling security measures that receive training data from multiple decentralized devices, and 6G's high-speed low-latency features enable real-time security patch deployment and model upgradings for quick threat responses [14]. Quick response times remain essential within healthcare because any slight postponement usually leads to negative patient results. Healthcare providers can establish adaptive security solutions through an integration of federated learning alongside 6G networks [15]. Intrusion detection algorithms receive regular updates derived from linked devices since cyber threats evolve. This makes the system exceptionally resistant to advanced security threats because it remains safe at all times. The distributed learning method enables organizations to maintain reasonably priced adaptive security protocols that protect against current cyber threats [16].

2.1 Communication and Network Challenges in 6G-Enabled IIoMT Systems

Implementing 6G networks in Industrial Internet of Medical Things (IIoMT) systems opens many opportunities for developing new approaches in the healthcare domain; however, it brings certain communication and network challenges. Data management is among the crucial issues in IIoMT systems for dense connections because the number of linked devices produces real-time medical data [16]. The highly speculated 6G bandwidth and latency requirements must ensure that the data exchange is efficient, and safely replaces traditional means, in specific sectors such as neurosurgery and intensive care units. Standardization, together with interoperability, represents major problems. Various protocols and specifications power the wide range of IoT devices that include large diagnostic systems and wearable health monitors. Multiple devices need a unified communication architecture for proper operation so they can function efficiently. Healthcare organizations face operational efficiency issues when they need to integrate devices because they do not follow standardized integration methods. Privacy as well as security stand out as major concerns. Private patient data exchanged through 6G networks makes IIoMT systems vulnerable to cyberspace threats. Security of sensitive data demands encryption methods supported by secure communication frameworks to stop breaches and unauthorized access. Security solutions of today must find the right equilibrium that protects both speed and data integrity. An additional challenge emerges from managing interference in the high-frequency spectrum of 6G. The speed benefits obtained by terahertz and millimeter-wave bands create more opportunities for signal interference, especially within healthcare environments rich in connections. Network reliability requires both efficient spectrum management and effective interference control systems because they maintain uninterrupted medical procedures. Energy efficiency stands as an immediate problem for IIoMT systems. Hospital settings often lack the feasibility to recharge battery-powered medical IoT devices because they depend on battery power. Device reliability and operational span depend heavily on creating low-energy communication methods that work in real-time but avoid creating vast electricity usage [3]. The deployment of IIoMT systems enabled by 6G faces additional challenges because of joining networks alongside regulatory requirements. Healthcare networks need to maintain connections between different manufacturer devices while respecting strict privacy regulations in healthcare information. The installation of safe data processing platforms must happen to ensure both legal compliance and maintenance of networking connections between medical devices. The combination of energy-efficient communication protocols with better network reliability techniques and federated learning models enhances the safety and effectiveness of 6G-enabled IIoMT systems. 6G technology will bring a more interconnected medical care future to the healthcare sector when these obstacles are overcome.

3 Model Aggregation and Update Strategies

The evolving landscape of the 6G-enabled IIoMT requires essential synchronization between model aggregation and update procedures to improve intrusion detection based on federated learning within the fast-changing environment. A combination of security and adaptability enhancements requires well-defined methods for effective update distribution since routine local IIoMT device training contributes to the global model development. Task anonymity requires protection while achieving optimal results between data transfer costs, processing efficiency, and knowledge precision. Weighted averaging stands as the preferred aggregation technique, which unites local models by monitoring their training data distribution across devices. The system ensures that devices providing bigger data collections will influence the final model much more strongly. Conventional approaches designed for uniformly spread data under steady network conditions experience failures because IIoMT devices have dynamic behaviors as well as limited resources. Under specific circumstances, the implementation of asynchronous model updates becomes necessary as devices can transmit their model modifications at unintegrated times. Online healthcare applications need these requirements first because they provide real-time model updates with reduced delays.

Personalized model aggregation shows potential as an approach that modifies global models to suit the individual requirements of IIoMT devices. This technique matches up-to-date models with the operational needs of various healthcare devices through its tailored model application. Secure aggregation algorithms protect model update integrity by stopping adversarial attacks, along with guaranteeing that one entity cannot misuse the model without permission. The integration of 6G networks enables more efficient model aggregation because linked devices achieve speedy speeds, minimal latency, and seamless model distribution. The system's security alongside patient data protection enables IIoMT systems to track down new security threats at all times. The protection of medical data during model updates is achieved by federated learning methods together with complex cryptographic techniques [3, 6].

Model aggregation techniques require continuous development in order to solve these problems. Future developments of the next-generation intelligent healthcare systems should concentrate on three essential areas, which include efficient communication protocols, adaptive learning frameworks, and privacy protection methods. Manufacturers using IIoMT have gained access to an improved medical ecosystem through the integration of state-of-the-art 6G technology with federated learning. Figure 1 presents the federated learning architecture as it demonstrates a decentralized machine learning technique that enhances security and privacy. Several data owners (A, B, and C) conduct model training at local levels using their datasets while shying away from sharing their raw data with a centralized server. After completing training, the model updates get transmitted, but the central aggregation server avoids receiving raw data. The aggregation server strengthens the global model through integrated updates without gaining access to any sensitive information. The aggregation server makes continuous learning across multiple decentralized sources possible by distributing the combined model back to all data owners after aggregation. Such a method delivers outstanding benefits to business domains that prioritize data protection through regulatory laws such as healthcare and banking sectors and IoT systems.

Federated learning delivers dual advantages through its approach to reduce data transmission costs, together with security threats. Furthermore, network congestion remains low because model changes form the basis of sharing while personal data remains protected locally. Through federated learning, users receive enhanced dataset adaptability with measures in place to maintain GDPR and HIPAA privacy regulations. Federated learning allows organizations to increase the effectiveness of their models by learning from scattered data in the organizations and achieving the potential of AI with privacy. This developing technology will produce important effects on building protective AI programs for diverse industries that use intelligent modeling.

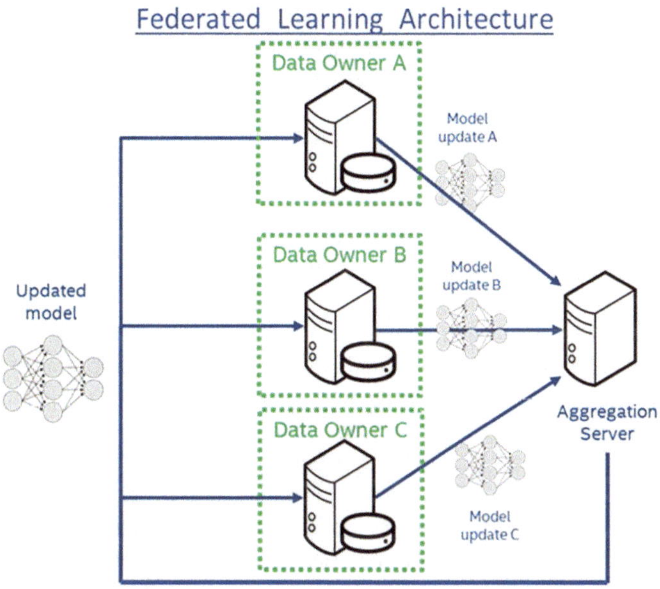

Fig. 1 Federated learning architecture

3.1 Performance Evaluation and Key Metrics in 6G-Enabled IIoMT Systems

Performance evaluation is critical for 6G-enabled IIoMT systems for federated learning-based intrusion detection to be effective, adaptive, and efficient. They are functionally endowed to provide medication and health upgrade devices in critical healthcare places where an outbreak of danger will depend on their abilities to detect threats, preserve privacy, and ensure uninterrupted reliability. Detection accuracy is one of the essential performance indicators that denotes how well the system recognizes legitimate and illegitimate activities. It is generally assessed through different metrics, including those of accuracy, precision, recall, and the F1- score, such that the system can correctly catch attacks with less frequent false alarms, which becomes very vital in a healthcare condition where missing a security threat could have detrimental consequences-the less number of false positives is also considered important because too many alarms might create overload situations for healthcare professionals, causing unnecessary distress and diminished system confidence (Fig. 2). In the context of 6G, the importance placed on real-time detection and reaction brings latency into consideration as a major factor. 6G application aims at enabling real-time security monitoring with an ultra-low latency to react to intrusions and model modification as quickly as possible. Whenever a threat needs to be identified and responded to, if the system has a longer latency, its efficiency relating to healthcare operations and patient safety would be at risk. Another such aspect of great importance is the communication overhead. It defines the amount of data transferred between IIoMT devices and their federated learning aggregate server. Reduce this communication overhead for the smooth functioning of IIoMT devices that normally cannot afford high processing and network resources. Efficient modeling updates tend to eliminate a lot of redundant traffic, bringing better performance to the broad network bandwidth, as well as system efficiency in general.

Another serious issue is scalability. Healthcare networks are continuously growing with new sensors and medical devices. Increasing numbers of devices will need to be supported by the system without compromising its efficiency. Scalability is measured by factors such as the number of training epochs to convergence and the time taken for training and updates. A properly designed architecture should very well support the applicant's growing demands without any perceptible degradation in performance. Last but not least, battery-powered devices account for a great part of IIoMT. All energy-efficiency aspects would help extend the battery lifespan, while security performance would still allow devices to function without frequent recharging, which is a concern in most remote and resource-constrained contexts. Federated learning and 6G connection have tremendous prospects for giving birth to a next-generation healthcare system that is secure, scalable, and effective by closely evaluating these important performance measures. This trend will, in turn, foster super-resilience and future-proofing for IIoMT systems as it augments the protection and preservation of patient data, increases threat detection, and guarantees the smooth delivery of medical services.

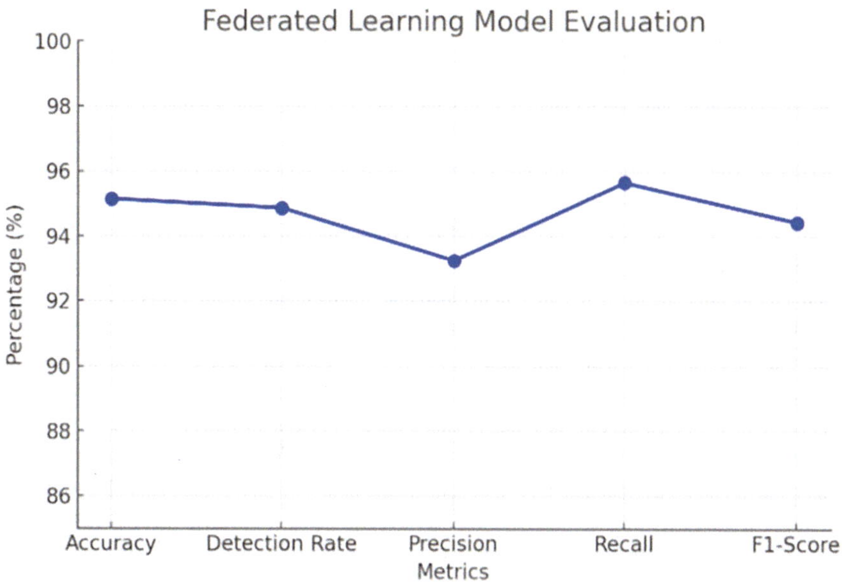

Fig. 2 Performance evaluation

Figure 3 shows how incoming communication gets monitored through network traffic analysis to detect dangerous activities. The security analysis confirms that a DDoS attack occurred with 98.7% certainty. To protect the system, it chose to block dangerous network traffic. After implementing the countermeasure, the system confirms the absence of remaining security threats. The benefits of IDS for cybersecurity become evident due to their ability to examine network patterns and identify abnormal behavior before taking security measures to defend the network infrastructure and maintain continuous operations.

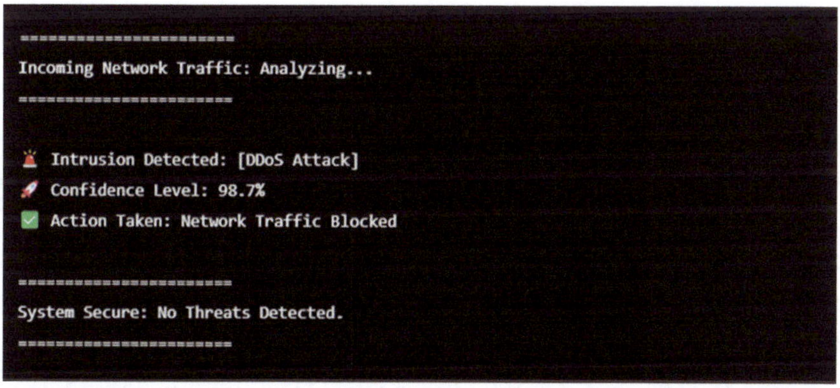

Fig. 3 Confidence level

3.2 Regulatory and Ethical Considerations in 6G-Enabled IIoMT Systems

As 6G-enabled IIoMT systems continue to transform the healthcare domain, ethical and legal issues abound. The sensitivity of patient data in networked medical devices engages these systems in legal frameworks like the GDPR in Europe and the HIPAA in the United States. A strong method of protecting privacy is federated learning, which minimizes any chance of a breach by keeping data decentralized. Still, only legal compliance does not address all ethical issues concerning IIoMT application. Algorithmic bias is one of the most prominent concerns. The author argues that if the data used to train machine learning models do not adequately represent all patient demographics, disparities in healthcare services may arise. In other words, a lack of diversity in training data may lead an intruder detection model to miss cyberthreats targeting certain pieces of equipment or hospital networks. Machine learning models must therefore be equitable and trustworthy for ethical AI in healthcare. Patient data ownership and consent are another key issue. For optimum security and diagnoses, IIoMT systems require constant data sharing, but patients still need to have control over their data. A full understanding and consent from the patient in all aspects of gathering, processing, and using medical data are of utmost importance. There are very important factors when making moral decisions, namely accountability and responsibility. For data security and regulatory compliance to be guaranteed, network providers, makers of medical devices, and healthcare organizations must work together. If any security mechanism, the system fails due to cyberattacks or system defects, it may jeopardize patient safety and reduce general public confidence in IIoMT technology. In addition, there are lingering concerns regarding possible health hazards linked with broadcast frequencies at very high ranges such as the 6G. Although research is still very much underway, some evaluation needs to be done about any long-term possible biological effects, as well as building safety standards for prolonged exposure.

4 Case Study: Federated Learning for Anomaly Detection in Smart Hospitals

The hospital group improved the security levels of its IIoMT equipment, including wearables, remote patient monitors, and infusion pumps, through the adoption of intrusion detection based on federated learning technology. Each device examined received data for anomalies as part of the method before sharing insights with the central system, which led to early cyber risk identification, including packet sniffing and DDoS attacks—the threats were accurately identified 96% of the time, which produced an improved security level for the hospital network [3]. Federated learning expanded security coverage to thousands of devices through a policy that protected patient information within the privacy regulations of HIPAA. The federated learning system operates differently from previous centralized methods because it prevents the transmission of sensitive patient information to achieve lower data breach vulnerability. Anomaly detection in real-time lets healthcare organizations fight cyber threats while maintaining medical procedures, so patient care can remain uninterrupted and of high quality. Enhanced security measures should protect IIoMT devices since they become increasingly networked. The hospital's federated learning approach proved that security protection could be utilized for effective operations, along with full compliance and efficiency maintenance. The method enhanced privacy while maintaining safe hospital infrastructure operations.

- As the medical network environment, along with all personnel involved in its maintenance and security, evolves, the challenge of protecting it from all sorts of threats has only grown. Such threats keep escalating, and now cybersecurity deals with one of the scapegoats in the line of such concerns—the sheer multiplication of methods of cyberattacks. The ever-increasing lure is directed toward secure clinical information and the control of life-preserving equipment, namely wearable medical devices, infusion pumps, and remote patient monitoring systems. Cybersecurity matters because even a small hack would have life-threatening consequences for the patients using these systems [17].
- A significant challenge lies beyond patient data privacy. Health confidentiality cannot withstand any breach; the traditional security methods need the collection of huge quantities of patient data in a centralized manner. This strategy raises enormous vulnerabilities to breaches of confidentiality and goes against HIPAA regulations. Triggering compliance with such security rules in a hospital environment, which contains thousands of connected devices, by human intervention is impossible. The autonomous monitoring of these activities, scalable concerning changes in network design and implementation, is a great need where human effort begins and ends to take care of the current IT workload while ensuring the integrity and confidentiality of information in the healthcare setting [18].
- Speed also plays a significant role in averting quantum threats. While the hackers plan and execute a complicated attack in seconds, conventional security technologies are simply not fast enough to identify the threat in real time. Without the benefit of fast detection of a probable threat, hospitals could be facing a difficult moment intervening in the patient, stopping system failure, or preventing data manipulation. The need of the hour is that healthcare units deploy proactive AI-oriented security so that they go one step ahead of the timely crimes. The measures must engage in immediate detection and elimination of threats before they wreak any possible havoc [19].

4.1 Federated Learning-Based Solution

The IDS training occurred locally within each IIoMT device using federated learning techniques at the hospital, which only distributed model updates rather than hospital data to its centralized FL server. The process included (Fig. 4):

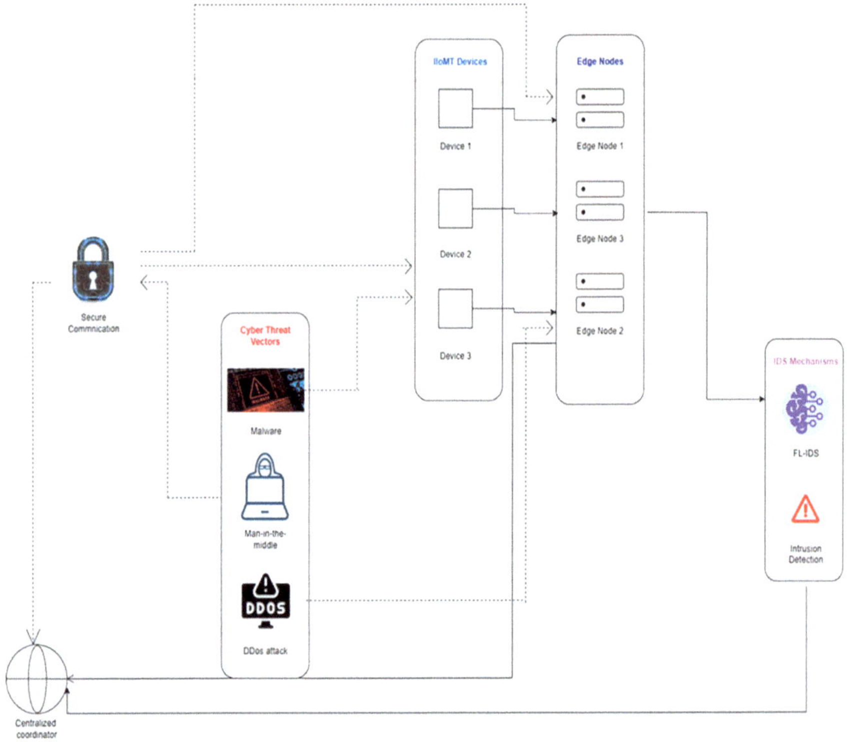

Fig. 4 Process of federated learning on IMoT devices

- **Local anomaly detection**—Every device monitored unusual incoming traffic patterns for irregularities.
- **Collaborative model training**—The federated learning system acquired encrypted model updates through its server from the connected devices.
- **Global model aggregation**—The central FL server operated to aggregate information that enhanced the global intrusion detection model.
- **Real-time anomaly detection**—The redesigned model received distribution among devices to enhance threat detection speed as well as precision.

The solution successfully detected all threatening attacks, including packet sniffing, DDoS attacks, and unauthorized access attempts, with 96% success. Data protection operates under the frameworks of GDPR and HIPAA by hosting patient information on-site to secure complete data privacy. As part of the plan, the system could protect 10,000 devices while remaining within the network's capacity threshold. The solution increased reaction speed by detecting cyber threats faster than standard centralized approaches when performing identification operations. Figure 5 describes the detection of cyberattacks in smart devices such as wearable monitors, infusion pumps, imaging scanners, bedside sensors, and smart ventilators.

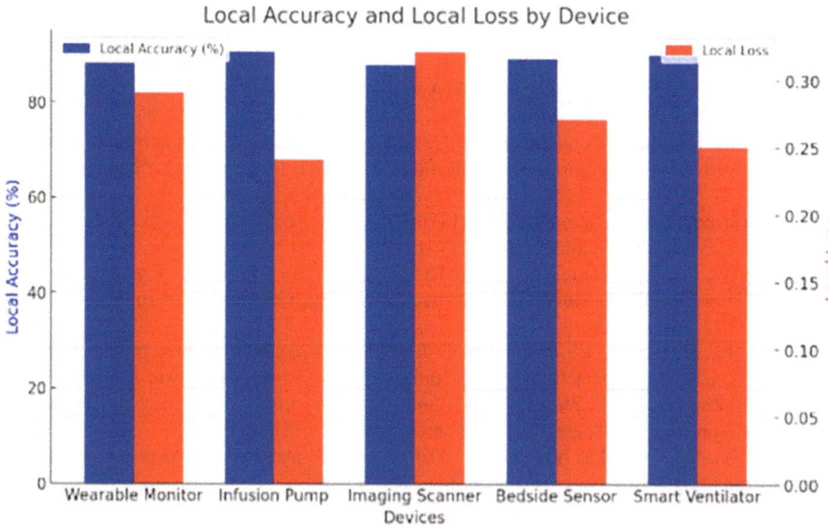

Fig. 5 Detection of cyberattacks on smart device

This system identifies the cyberattacks of ransomware, man-in-the-middle attacks, distributed denial of service, and botnets.

5 Conclusion

Federated learning concerning securing IIoMT systems in 6G, federated learning has proven to be a revolutionary concept. With the new decentralized security implementation, healthcare entities were able to protect, comply with healthcare standards, and prevent the compromise of patient data privacy with thousands of smart medical devices, such as wearables, infusion pumps, and remote monitoring systems, from cyber-threats that were in a constant state of evolution. The approach, which enhanced operational efficiency and ultimately improved data security, gave healthcare workers the peace of mind to focus on patient-related activities and not worry about the incessant cyberattack threats. Very significantly, by not allowing security protocols to interfere with critical medical workflows, Medica assisted in protecting its application in the hospital environment. In the time to come, federated learning, along with 6G technologies, will be steering healthcare cybersecurity extensively, providing scalable, real-time threat detection and guaranteeing uninterrupted and secure medical services. Hospitals can stay compliant with the regulations and stay abreast with the ever-changing cybersecurity threats by prioritizing privacy-preserving AI technologies. Ultimately, we can say that federated learning is a strong punch for keeping patients safe and imposing faith on frontline medical technology.

References

1. Madanian S, Chinbat T, Subasinghage M, Airehrour D, Hassandoust F, Yongchareon S (2024) Health IoT threats: survey of risks and vulnerabilities. Fut Internet 16(11):389
2. Baucas MJ, Spachos P, Plataniotis KN (2023) Federated learning and blockchain-enabled fog-IoT platform for wearables in predictive healthcare. IEEE Trans Comput Soc Syst 10(4):1732–1741
3. Abdel Hakeem SA, Hussein HH, Kim H (2022) Security requirements and challenges of 6G technologies and applications. Sensors 22(5):1969
4. Nguyen VL, Lin PC, Cheng BC, Hwang RH, Lin YD (2021) Security and privacy for 6G: a survey on prospective technologies and challenges. IEEE Commun Surv Tutor 23(4):2384–2428
5. Ferrag MA, Maglaras L, Moschoyiannis S, Janicke H (2020) Deep learning for cyber security intrusion detection: approaches, datasets, and comparative study. J Inf Secur Appl 50:102419
6. Wang M, Zhu T, Zhang T, Zhang J, Yu S, Zhou W (2020) Security and privacy in 6G networks: new areas and new challenges. Digit Commun Netw 6(3):281–291
7. Li T, Sahu AK, Talwalkar A, Smith V (2020) Federated learning: challenges, methods, and future directions. IEEE Signal Process Mag 37(3):50–60
8. Abbas SR, Abbas Z, Zahir A, Lee SW (2024) Federated learning in smart healthcare: a comprehensive review on privacy, security, and predictive analytics with IoT integration. In: Healthcare, vol 12, no 24, p 2587). MDPI
9. Rehman A, Abbas S, Khan MA, Ghazal TM, Adnan KM, Mosavi A (2022) A secure healthcare 5.0 system based on blockchain technology entangled with federated learning technique. Comput Biol Med 150:106019
10. Shahraki A, Abbasi M, Piran MJ, Taherkordi A (2021) A comprehensive survey on 6G networks: applications, core services, enabling technologies, and future challenges. arXiv:2101.12475
11. Li B, Wu Y, Song J, Lu R, Li T, Zhao L (2020) DeepFed: federated deep learning for intrusion detection in industrial cyber–physical systems. IEEE Trans Industr Inf 17(8):5615–5624
12. Popoola SI, Ande R, Adebisi B, Gui G, Hammoudeh M, Jogunola O (2021) Federated deep learning for zero-day botnet attack detection in IoT-edge devices. IEEE Internet Things J 9(5):3930–3944
13. Al-Huthaifi R, Li T, Huang W, Gu J, Li C (2023) Federated learning in smart cities: privacy and security survey. Inf Sci 632:833–857
14. Mothukuri V, Khare P, Parizi RM, Pouriyeh S, Dehghantanha A, Srivastava G (2021) Federated-learning-based anomaly detection for IoT security attacks. IEEE Internet Things J 9(4):2545–2554
15. Aledhari M, Razzak R, Parizi RM, Saeed F (2020) Federated learning: a survey on enabling technologies, protocols, and applications. IEEE Access 8:140699–140725
16. Zhao R, Gui G, Xue Z, Yin J, Ohtsuki T, Adebisi B, Gacanin H (2021) A novel intrusion detection method based on lightweight neural network for internet of things. IEEE Internet Things J 9(12):9960–9972
17. Ruzafa-Alcázar P, Fernández-Saura P, Mármol-Campos E, González-Vidal A, Hernández-Ramos JL, Bernal-Bernabe J, Skarmeta AF (2021) Intrusion detection based on privacy-preserving federated learning for the industrial IoT. IEEE Trans Industr Inf 19(2):1145–1154
18. Ahad A, Tahir M, Yau KLA (2019) 5G-based smart healthcare network: architecture, taxonomy, challenges and future research directions. IEEE Access 7:100747–100762
19. Lu Y, Huang X, Dai Y, Maharjan S, Zhang Y (2020) Federated learning for data privacy preservation in vehicular cyber-physical systems. IEEE Netw 34(3):50–56
20. Rehman A, Razzak I, Xu G (2022) Federated learning for privacy preservation of healthcare data from smartphone-based side-channel attacks. IEEE J Biomed Health Inform 27(2):684–690
21. Arikumar KS, Prathiba SB, Alazab M, Gadekallu TR, Pandya S, Khan JM, Moorthy RS (2022) FL-PMI: federated learning-based person movement identification through wearable devices in smart healthcare systems. Sensors 22(4):1377

22. Sikandar HS, Waheed H, Tahir S, Malik SU, Rafique W (2023) A detailed survey on federated learning attacks and defenses. Electronics 12(2):260
23. Liu Y, Yuan X, Xiong Z, Kang J, Wang X, Niyato D (2020) Federated learning for 6G communications: challenges, methods, and future directions. China Commun 17(9):105–118

Blockchain-Powered Intellectual Property: A Decentralized Approach to IP Management

S. Madumidha, G. Abhishek, M. Akilesh Krishnan, K. J. Hemaharshini, P. Sivaranjani, K. Suresh Kumar, and Sunday Adeola Ajagbe

Abstract The Proof-of-Contribution framework will serve as the foundation for the first blockchain-enabled web application specifically designed to provide enhanced protection for intellectual property (IP). Most existing IP systems, including patents and copyrights, are rife with inefficiency, nontransparent, and cannot withstand false claims, which damage the rights of innovators. This application will provide record-keeping for contributions made and rights verification in a transparent and immutable ledger through the application of blockchain technology. All forms of intellectual contributions and patents can be uploaded by users, whereupon the system will conduct cross-verification of submissions against existing documents to determine duplication or unauthorized claims against another patent. After verification, entries will be on the blockchain, securing their persistence (including date-time stamp, contributor details, and ownership metadata). Through this solution, trust will be enhanced, disputes minimized, and attainment of a fair patent filing and validation process that will eventually improve the attribution of innovations will be achieved.

Keywords Blockchain · Intellectual property · Proof-of-contribution · Patent protection · Immutable ledger · Contribution verification · Transparency

S. Madumidha · G. Abhishek · M. A. Krishnan · K. J. Hemaharshini
Department of Information Technology, Sri Krishna College of Technology, Coimbatore, Tamil Nadu, India

P. Sivaranjani
Department of ECE, Kongu Engineering College, Erode, Tamil Nadu, India

K. S. Kumar (✉)
Department of Computer Science and Engineering, Sri Eshwar College of Engineering, Coimbatore, Tamil Nadu, India
e-mail: sureshkumarkskphd@gmail.com

S. A. Ajagbe
Computer Engineering Department, Ladoke Akintola University of Technology LAUTECH, Ogbomoso, Nigeria

A. Kumar et al. (eds.), *Blockchain and Federated Learning Synergy for Privacy-Focused DeepFex Solutions*, Blockchain Technologies,
https://doi.org/10.1007/978-981-95-1394-9_10

1 Introduction

Intellectual property (IP) safeguarding is crucial for protecting creative works from infringement, unauthorized use, or exploitation. Historically, their legal instruments have been patents, copyrights, and trademarks. However, with the digitization of many activities, those instruments have been seen to suffer some challenges, including taking much time to process, cost considerations of legal procedures, and difficulties in establishing the origin or ownership of ideas [1]. As data and ideas travel more rapidly and globally, there comes a definite need for a faster and clearer means of ensuring that innovators get recognition and protection for his or her contributions.

Blockchain technology is perceived nowadays as an excellent antidote to all those limitations imposed by traditional IP protections. Thus, the above-mentioned features of a decentralized, tamper-proof ledger system on which intellectual contributions can be securely, traceably, and immutably recorded can be viewed as a strong platform for IP preservation [2]. It also reduces such disputes by letting them ensure that the work is timestamped and verifiable, hence ownership and originality minimization. This paper attempts to look into how blockchain would eventually change IP protection through the lens of the Proof-of-Contribution (PoC) scheme within a digital environment.

1.1 Blockchain

Utilizing blockchain for IP management offers a multitude of significant benefits that address the shortcomings of traditional systems. One of the primary advantages is enhanced security. Blockchain technology provides a tamper-proof record of ownership and transactions, ensuring that creators can confidently assert their rights over their IP [3]. Each transaction is cryptographically linked to the previous one, creating an immutable chain of records that is resistant to alteration or fraud. This level of security is particularly crucial in an era where digital content is easily copied and distributed, as it helps protect creators from unauthorized use and infringement.

In addition to security, blockchain enhances transparency in the management of IP. The decentralized nature of blockchain allows for real-time tracking of IP assets, making it easier for creators, licensors, and users to verify ownership and usage rights [4]. This transparency can significantly reduce disputes over IP ownership, as all transactions are recorded on a public ledger that is accessible to authorized parties. By providing a clear and verifiable history of ownership, blockchain can help resolve conflicts more efficiently and fairly, fostering trust among creators and users alike.

Another critical benefit of using blockchain for IP management is the potential for cost and time savings. Traditional IP registration processes can be cumbersome, involving lengthy paperwork, high legal fees, and the need for intermediaries such as lawyers and administrative bodies [5]. Blockchain can streamline these workflows by automating tasks through smart contracts—self-executing contracts with the terms of the agreement directly written into code. This automation reduces the administrative burden on creators and allows for faster processing of IP registrations and licensing agreements. As a result, creators can focus more on their work rather than navigating complex legal frameworks.

Moreover, the decentralized nature of blockchain fosters a more collaborative environment for creators. By enabling peer-to-peer interactions, blockchain allows creators to share and license their work more easily while maintaining control over their IP. This capability is particularly beneficial in creative industries where collaboration is essential, as it encourages innovation and the sharing of ideas without the fear of losing ownership [6]. Creators can set specific terms for how their work can be used, ensuring that they receive appropriate compensation and recognition. Additionally, blockchain can facilitate the creation of new business models for monetizing IP. For instance, creators can tokenize their work, allowing them to sell fractional ownership or license their IP in a more flexible manner [7]. This opens up new revenue streams and opportunities for creators, particularly in the digital realm where traditional monetization methods may be less effective.

Finally, the global nature of blockchain technology means that it can provide a standardized approach to IP management across borders. This is particularly important in an increasingly interconnected world where ideas and innovations cross-geographical boundaries [8]. By offering a universal platform for IP registration and management, blockchain can help harmonize IP laws and practices, making it easier for creators to protect their rights internationally.

2 Literature Review

In an ever-growing, competitive digital environment, traditional IP systems are met with numerous obstructions to validation and enforcement. The existing literature review will thus provide an analysis of current systems' inefficiencies, effects of the digital environment, and the emerging technologies like blockchain-based possible remedies [7].

The cumbersome nature of traditional IP systems is aggravated by delays in the granting of patents and registrations of copyright. Centralization means patent offices usually have huge backlogs of applications and, hence, long delays. In the United States, the entire process of obtaining a patent, depending on the complexity of the invention and the resources of the patent office, may take somewhere between two to five years on average. Such delays hinder commercialization, thereby increasing the risk of legal confrontation, since competitors might infringe upon an idea before its formal protection [9].

Kingston also emphasizes that these delays inhibit technological development and hinder market competition by benefiting large corporations that have the advantage of pursuing legal action in case of infringement [10]. It can be too expensive for many small enterprises and individual creators to acquire IP rights and enforce their rights in the event of infringement. Often, the costs associated with patent filings, maintenance fees, and legal disputes mount like an avalanche, leaving many creators unable to ensure even minimal protection. Patent filing costs in the United States may thus range from $5,000 to $15,000, with litigation expenses mostly exceeding $1 million. Huge legal costs in the domain of IP effectively shut out smaller innovators and deny them recourse to defend their rights [11]. According the findings proposed by Carlson [12] states that inefficiencies are aggravated by a lack of transparency in traditional IP systems, where unclarified ownership records and frustrating bureaucratic red tape constitute further pitfalls to creators [12].

Ownership attribution for invented or creative work becomes much more complicated, with interdisciplinary and collaborative research. In the sectors of biotechnology or software development, several teams usually contribute to creating a product. A centralized IP system does not possess the ability to deal with this complexity, and this leads to disputes over ownership and contribution [13]. According to Zyskind et al., decentralization of privacy through blockchain can also lessen these disputes since it will keep verifiable and immutable records of contributions [14]. Most IP laws are national or regional, meaning that enforcement across borders relies on World Intellectual Property Organization (WIPO) treaties [15]. Some of the countries have weak mechanisms for enforcement, which makes it difficult to enforce rights globally. The authors highlighted the problems of challenging global enforcement and mentioned that blockchain technology might come to introduce a new paradigm for cross-border IP protection.

In addition, the emergence of digital platforms has changed the scene of IP management. Barman [16] showed that platforms such as YouTube and Instagram enable the sharing of content and show efforts to enforce copyright on that content due to the sheer amount of user-generated content [16]. According to the authors smart contracts and NFTs hold the awesome possibility of refashioning royalty payment principles and creating a new manner for content distribution, fair to the creator.

Piracy has been a predominant nemesis of industries like films, music, and software. Such is the scenario envisioned by Yang [17], who explains the role of blockchain in IP management by virtue of platforms such as IPwe, whose mission is to simplify patent transactions and enforcement [17]. Zohar [23] also explores the principles underlying Bitcoin, showing how the tamper-proof nature of blockchain can be leveraged for IP rights [13]. As indicated in its whitepaper, Flow blockchain offers a scalable and highly efficient approach to managing digital ownership, especially around NFTs and tokenized assets; thus, making it an important instrument in IP protection [9].

The emergence of blockchain has proven to be a promising solution to several problems of traditional IP systems. The introduction of the original idea of blockchain through Bitcoin was done by Aveni [18]. He explained how decentralized records could improve security and transparency [18]. Bogdanov [19] puzzles out that the blueprint of a new economy where permanent account of digital ownership exists is the blueprint for blockchain [19]. According to Buterin, Ethereum's smart contracts serve as a mechanism for automated management of IP rights via decentralized applications [20]. Enterprises, according to Mougayar, will evolve in an open blockchain environment, which digitally enhances property rights through programmable ownership structures [21].

Smart contracts have changed the scenario completely in which IP rights are managed and enforced. This is because these contracts are executed automatically when certain predefined conditions are met. Whenever certain conditions are satisfied, smart contracts execute their process automatically. The initial processing for the licensing of IP generation may be replaced by such contracts where the payer is sure that the creator will be compensated for their work by getting everything done automatically when the conditions are fulfilled [22].

Zohar [23] have explored many areas of application for smart contracts, especially their ability to substitute traditional licensing with automated, trustless agreements [23]. The WIPO also recognizes the application of blockchain in the enforcement of IP rights, stating how distributed ledger technology is apt to enhance tracking and resolution of disputes in IP cases [24].

Flow blockchain is one such emerging technology specifically designed as a highly scalable and developer-friendly environment for decentralized applications. Designed with an architecture that promotes high-throughput and very low-cost transactions, it can be an ideal choice for IP management [25].

Mentsiev [26] and El Koshiry [27] advocate for clear definition and understanding of regulations in the enforcement of rights as recorded on a blockchain by discussing the legal implications of IP management through blockchain [26, 27]. The comprehensive guide offered by [26] explains the said process as regards how blockchain will integrate with existing IP laws and rogues for seamless implementation.

As the concept of the Internet of Things (IoT) proliferated globally within intricate ecosystems, issues regarding data security and privacy became apparent [28]. Blockchains have many benefits, like better security, privacy, more space, and the ability to connect to other computers [29]. Before blockchain technology can be used to solve IoT problems, it needs to be fixed in many ways. Despite the perception of smart contracts as the forthcoming advancement in blockchain technology, numerous challenges remain to be resolved. The limitations and diversity of IoT must be taken into account when executing smart contracts. Moreover, engaging with smart contracts requires dependence on oracles that provide reliable real-world data.

3 Existing System (Lolite)

In the contemporary landscape of IP management, traditional legal instruments such as patents, copyrights, and trademarks have long served as the cornerstone for protecting the rights of creators and innovators. These mechanisms have been instrumental in establishing a framework for ownership and originality, allowing individuals and organizations to safeguard their inventions, artistic works, and brand identities [30]. However, as we navigate the complexities of the digital age, it becomes increasingly evident that these conventional methods are grappling with significant challenges that hinder their effectiveness.

Lolite is a comprehensive IP management ERP system designed to streamline the operations of IP attorneys, owners, advocates, and solicitors by facilitating a paperless workflow. This existing system effectively manages the entire workflow for Intellectual Asset Managers, reducing errors and ensuring that critical deadlines are met. It allows users to monitor the current and pending statuses of cases, including hearing dates, submission deadlines, and client communications, thereby enhancing the organization and efficiency of daily operations across various IP departments [31]. The software encompasses complete case workflow management for multiple IP sections, including patents, trademarks, designs, copyrights, litigation, and the Intellectual Property Appellate Board (IPAB). Key features include work allocation to staff, cover letter management, form management, hearing tracking, order management, billing, and client document organization. It also provides modules for managing specific case types, such as patent and trademark applications, opposition cases, and copyright registrations, ensuring comprehensive oversight and tracking of all relevant details [22].

The existing system is designed to facilitate the management of various aspects of IP, including case details, fee receipt management, renewal reminders, and workflow validation. It supports the organization of documents and the tracking of legal proceedings, making it a valuable tool for IP professionals. However, the proposed system aims to enhance these capabilities further by integrating blockchain technology, which promises to improve security, transparency, and efficiency in managing IP assets. In addition to the time-consuming nature of traditional IP processes, the financial burden associated with legal procedures cannot be overlooked [32]. The costs of filing patents, securing copyrights, and defending trademarks can be prohibitively expensive, particularly for individual creators and small businesses. Legal fees, application costs, and the need for expert consultations can quickly accumulate, creating a barrier to entry for many who wish to protect their IP. This financial strain can lead to a situation where only those with substantial resources can afford to safeguard their innovations, leaving a significant portion of the creative community vulnerable to infringement and exploitation. The critical challenge lies in the difficulty of establishing the origin or ownership of ideas.

In a world where information is shared instantaneously and globally, disputes over ownership and originality have become increasingly common. The traditional legal frameworks often struggle to address these disputes effectively, leading to protracted legal battles that can drain both time and resources. The ambiguity surrounding ownership can create an environment of mistrust among creators, further complicating collaboration and innovation. The complexity of existing legal frameworks adds another layer of difficulty. Navigating the intricacies of IP law often requires the expertise of lawyers and administrative bodies, which can be both time consuming and costly. This reliance on intermediaries not only increases the overall expenses associated with IP management but also introduces additional points of failure in the process.

The need for intermediaries can slow down the resolution of disputes and complicate the enforcement of rights, leaving creators in a precarious position. As the pace of innovation accelerates and the ease of information sharing continues to rise, it is clear that traditional methods of IP protection are struggling to keep up. The rapid exchange of ideas in today's globalized world presents unique challenges that demand more efficient and effective solutions. Innovators and creators are increasingly seeking alternatives that can provide them with the security and assurance they need to protect their work without the burdensome delays and costs associated with traditional systems.

4 Proposed System (Blockchain-Based IP Management)

Blockchain is a distributed ledger technology that records transactions in a secure, immutable, and transparent manner. It operates through a network of nodes that validate and verify transactions using consensus algorithms. Once validated, these transactions are recorded into blocks, which are then linked together in a chain, ensuring that no information can be altered without consensus from the entire network.

The PoC mechanism is a novel approach to IP protection wherein each participant's contribution is validated and recorded on the blockchain. Unlike traditional methods, where the creator files for protection, PoC ensures that all collaborators in a project receive credit for their work, particularly useful in research and innovation, where multiple parties often contribute to a single IP asset.

Another important algorithm is actually a consensus algorithm that allows decentralized nodes to reach consensus on the actual state of the blockchain without a central authority. Typical consensus algorithms include Proof of Work (PoW), which Bitcoin uses, and Proof of Stake (PoS), which Ethereum 2.0 and other blockchains use. Whereas PoW asks miners to solve hard mathematical problems, PoS chooses its validators based on the number of tokens that they hold and are willing to "stake" as collateral. These mechanisms help in preventing fraud and double-spending by including only valid transactions in the blockchain.

Public-key cryptography also comes into play with blockchain projects by ensuring transaction security, as well as maintaining users' authenticity. Some algorithms like Elliptic Curve Digital Signature Algorithm (ECDSA) are used to generate digital signatures to prove that a transaction has been authorized by the rightful owner of the private key. The hashing algorithm is one of the many cryptographic functions. Input (or message) data are used as parameters to produce a fixed-length character string, generally a concatenated series of numbers and letters, which is called the hash value or hash code. Given that the hash value output is unique for any input of data, hashing is mainly concerned with representing data in a small form so that, with a minute change in input, a drastically different output is produced as a hash value.

Applications for hashing algorithms involve data retrieval, password storage, digital signatures, and blockchain technology. They usually guarantee data integrity by allowing quick ways to check whether any data has ever been modified. One of the main features of a good hashing algorithm that ensures security is that it is computationally infeasible to get the original input just by getting the hash value. Some of the more popular hashing algorithms are SHA-256 (Secure Hash Algorithm 256-bit), MD5, and RIPEMD-160. In a blockchain, for instance, hashing would be crucial in forming blocks, making sure that data contained in them has validity, and securing the linking of those blocks in chain.

In conclusion, hashing algorithms provide a means of uniquely identifying data through an output of fixed length. It is a well-known and largely used concept in many areas, especially in verifying data and security.

Figure 1 describes the authentication process step by step, which works in the authentication layer.

(1) Registration: Either the user registers if he/she is a new user.
(2) Login: If the user has already signed up, then he/she can use the Key to login to the application.
(3) The login details are shared with the service provider, which checks with the blockchain to identify the user.
(4) The service provider checks the identity in the identity list using queries by the use of secret key entered by the user. It uses hashing for identity verification.
(5) If the user exists, it responds with the authentication request, which includes timestamp and other credentials.

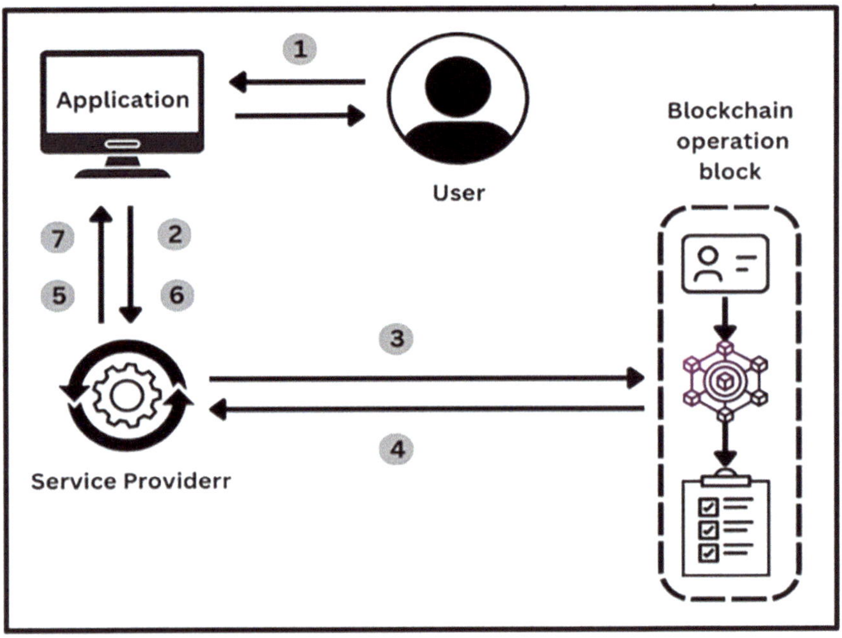

Fig. 1 Decentralized authentication process

(6) The s p passes the authentication request to the user through the application.
(7) Then, the user will create a signature with all the details, including username, public key, date, along with the timestamp.
(8) The service provider will verify the received information, and if the credentials are valid, then it will accept, else it will reject the user with login denied.

5 Methodology

An Internet-based application that uses blockchain to improve the management and protection of IP is the objective. It does this with the use of a PoC mechanism, which also helps to address some specific challenges that traditional IP management systems encounter, specifically transparency, security, and efficiency. The application centers around three major modules: contribution tracking, validation, and dispute resolution. Each is instrumental in ensuring that contributions are recorded, verified, and substantiated.

The first of these modules, contribution tracking, gives the system its foundation. Here, users—who may be individual creators, developers, or organizations—submit their works onto the blockchain. Each submission is complemented by metadata that provides information such as the identity of the creator, the nature of the contribution, and the date and time of submission. The contribution is thus timestamped and imprinted onto the blockchain so that a permanent and immutable record exists as to the contribution.

Since the record is held in a decentralized manner, all the stakeholders, including other contributors, project managers, and potential users of the IP, will have access to it. Therefore, granting a transparent view of contributions helps build trust among participants and minimizes the risk of disputes that can arise over ownership and originality.

The third module is about dispute resolution, which is the inevitable conflict that may appear in collaborative projects, especially in open-source software development, where several contributors come into play. Disputes in traditional IP management systems are long, tedious, and costly, killing innovations and collaboration along the way. On the contrary, the blockchain-based system creates a clear and immutable record of all contributions, thus simplifying conflict resolution. During a dispute, stakeholders will be looking into the timeline of contributions and the associated metadata on the blockchain for verification purposes.

Such transparency makes it possible to resolve disputes with relative ease, as evidence is readily available and indisputable. This facilitates quick and cheap conflict resolution, which, in turn, will be an incentive for greater collaboration and reduced friction in multi-contributor environments.

The advantages of this blockchain-based approach find maximum application in open-source software projects, where the collaborative nature of development often leads to complex contribution dynamics. Real-time tracking of contributions ensures that developers are recognized for their efforts, a factor that nourishes feelings of

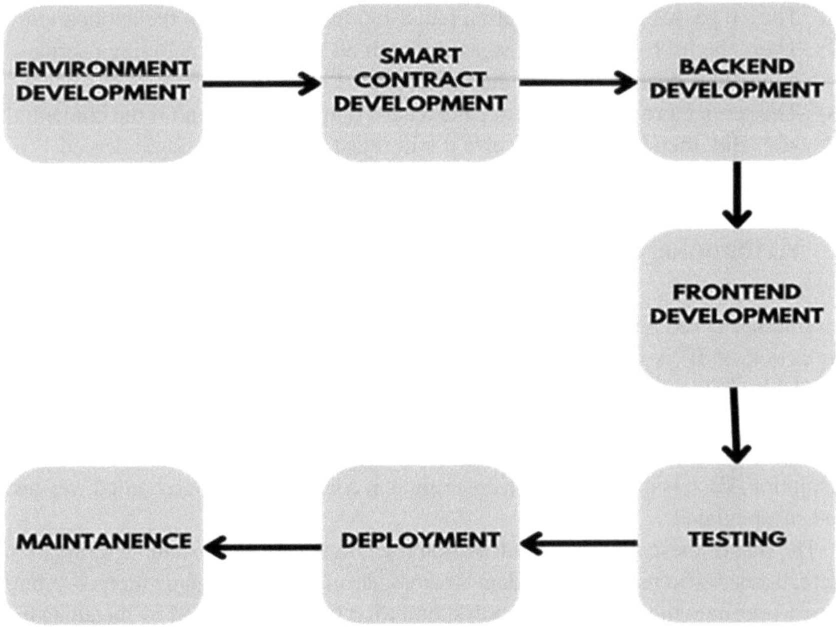

Fig. 2 Development process

ownership and motivation among contributors. Such recognition is crucial within open-source communities, where contributors often use reputation and acknowledgment to build their professional profiles. In this way, the extent of security that blockchain gives will also be embedded in the system. The development process is displayed in Fig. 2.

The typical development process for blockchain-based applications is given as a flow diagram, which describes the start with creating the application's core functionality, followed by developing the user interface. After testing and ensuring the application's readiness, it's deployed to a production environment. Finally, maintenance is ongoing to address any issues and keep the application updated.

5.1 Smart Contract Development

The development of smart contracts defines the processes during which one or more self-executing contracts are created, with terms of the agreement being expressed in lines of code. They are designed to independently hold and execute certain actions once predefined conditions have been met, thus eliminating the need for intermediates like lawyers and banks, on any issue pertaining to smart contracts. Smart contracts apply conditions on a blockchain, which is actually a safe, decentralized, and transparent environment for them to operate in.

Typically, smart contract development includes steps like:

Requirements gathering and planning stage: Developers generally work along with different stakeholders (e.g., businesses, legal practitioners, or project owners) in defining any specification before coding begins. This could consist of actions like transferring assets, managing users' rights, or verifying considerations of agreements like payments or transfer of ownership.

Code writing: Smart contracts are written by using programming languages that are purposefully created for every specific blockchain platform. For example, Solidity is used for Ethereum, while Cadence is used for flow blockchain. The logic of the contract will be encoded in these languages, describing the conditions to be met and the actions to be taken once they have been fulfilled. For instance, in a token transfer contract, conditions would be defined for the amount of tokens being transferred and the recipient's address.

Security audit: Smart contracts become benign and immutable once they are deployed on the blockchain, which means that a bug or vulnerability will cause very serious financial loss or some kind of legal issue. As with many other elements in this field, security audit will be a very important activity within which code is inspected either manually or automatically, to identify possible vulnerabilities, including reentrancy attack, overflow/underflow issues, and improper access controls. Usually, this step is forced by an external third-party security auditor for the contract to be considered secure.

Deployment: After finishing all these steps with much tougher tests and audits, the contract is ready for deployment over the blockchain. Deployment means submitting the smart contract code on blockchain network, making it immutable. After that, the contract comes into public view and would be open for access and transaction with users under the rules and conditions defined therein. This approaches most developers in choosing tools they would use, depending on the blockchain, Flow CLI, Truffle, or Remix for Ethereum. Thus ensuring that IP operates smoothly and securely, efficiently managing the IP assets.

5.2 *Backend Development*

Backend development refers to the creation of server-side components of the application, which deal with things such as processing the business logic, managing the data, and also help in communication between the frontend and the database. The backend thus focuses more on the server-side coding, which the servers run to handle user requests, process the data, and send responses. Also, the backend has to take care of all the communications with the databases regarding storing and retrieving data like that stored in user records, content, or transaction records. It secures and also makes sure that the application is efficient and fast by handling the authentication of users, validating data, and enforcing security protocols for the application. A backend development right now would surely use programming languages such as Python, Java, Ruby, or Node.js and depend on databases such as MySQL, PostgreSQL, or MongoDB for storage and management of data. This side of development mainly accounts for making sure that the core functions of the application work as intended, without performance, security, and scalability.

5.3 *Frontend Development*

The frontend development is the area responsible for the design and realization of the visible and interactive aspects of a website or application with which the end-user interacts directly. It deals with the design and development of the UI and UX in a way that is attractive and easy to use. Frontend development provides the look, feel, and interactivity of an application with layouts and structures on web pages. The interactivity enables the functionality of the application. It sets the architectural fittings in HTML, CSS, and JavaScript for developing responsive and dynamic web pages with many kinds of layouts, devices, screen sizes, and so on. Frontend developers use tools such as React, Angular, and Vue.js in order to optimize development toward a better user experience. In doing so, they ensure that the users may perform all their tasks smoothly and intuitively in the application. Whereas frontend and backend comprise the extremities of the technological domain, they apply the utmost beauty and adeptness in turning complex processes and data into a simple, usable interface. The home page is illustrated in Fig. 3.

The image in Fig. 3 is the landing page of the application, where the users will be directed upon hitting the application. This page shows the Navigation bar with options like Create, Explore, Github, and User Login.

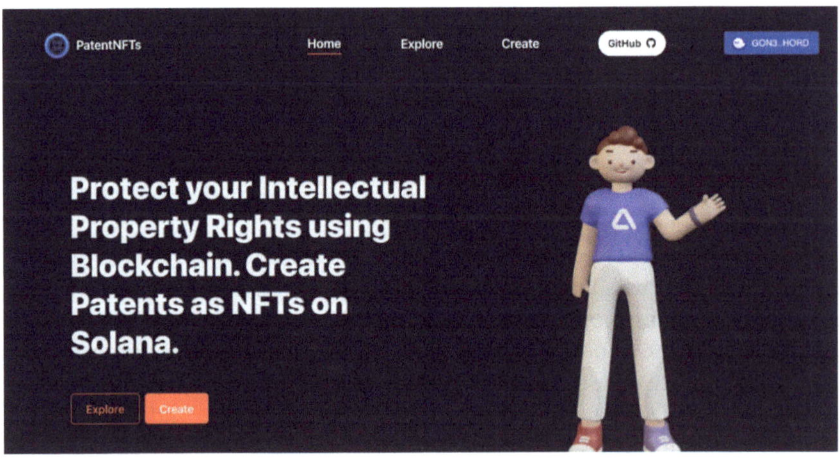

Fig. 3 Home page

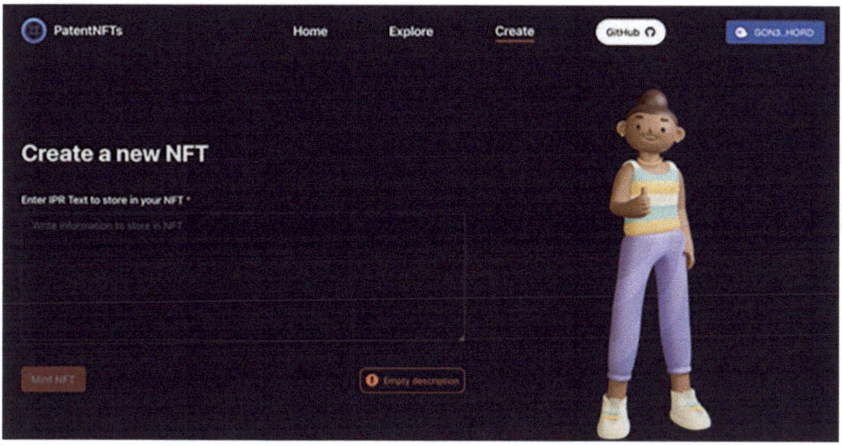

Fig. 4 NFT creation page

The NFT creation page (Fig. 4) of a blockchain project allows users to create, mint, and manage their own NFTs directly from the platform. It has a straightforward user interface that allows the user to upload a digital asset, be it an image, video, music, or even a 3D model, that the users want to mint into an NFT. The user then enters information, including the name, description of the asset, and any specific attributes or properties for imprintation for the NFT, such as whether or not it will be rare or have special traits. After this full description entry by a user, the back-end interface of the page connects to the blockchain, mostly by using smart contracts, in order to mint an NFT. A smart contract associates the NFT with a wallet address of the owner's entry, ensures that the NFT is unique, and stores metadata of the digital asset's file, information of creator, and historical transactions into a secure blockchain.

Figure 5 shows the page that is viewed by the user who uploads the documents to the blockchain for security. After uploading the file, they can find the uploaded document with the name, owner wallet, token generated, and the description that is given along with that.

Figure 6 shows the list of patents that were filed. It shows the description and the date of upload for every user who logged in. This page is also to explore what the other users shared on the website. It is like a data list where all the patents are available to search and look into.

5.4 Testing

The many testing processes have been conducted by a blockchain project so as to secure and validate, and thus ensure proper functioning of the entire system. The first

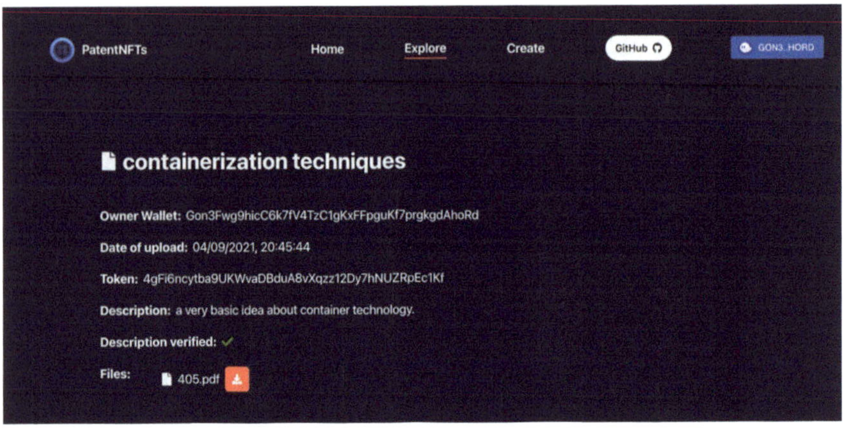

Fig. 5 Explore page

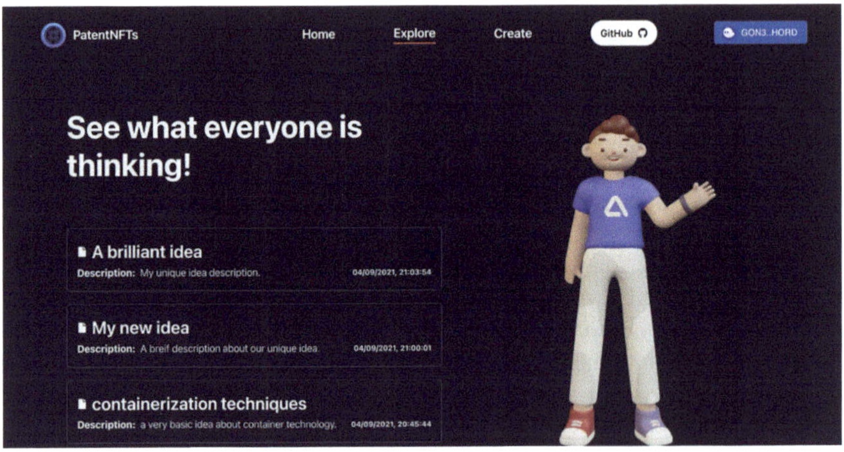

Fig. 6 List page

step would be writing a test plan and clarifying its focus areas. Such areas are usually logical with regard to smart contracts, networking functionality, security, and performance. Additionally, it would include conditions under which those smart contracts were expected to operate correctly, whereas the relevant behaviors on the blockchain network are defined—block generation, consensus, and the scaling capability of such a system, which supports high transaction throughput.

Unit testing is an essential testing method for any blockchain project. If your project contains smart contracts, unit testing establishes whether the functions in the smart contracts work the way they are supposed to. Tools like Truffle, Hardhat, or Brownie are applicable since they provide the frameworks for writing unit tests. Usually, such tests include valid and invalid input to effectively ensure that the smart contracts are resistant to such potential attacks. In Hardhat, for example, you can test whether the contract deploys correctly and that you receive expected values when interacting with that contract.

Then integration testing comes after unit testing and integrates into the chain testing with its components. Some of the things being tested here would be whether an interrelationship between accounts works, whether the correct validation of blocks is being done, and whether the consensus mechanism would behave normally. Integration testing helps ensure that the blockchain network and smart contracts work together perfectly, thus creating a perfect, holistic representation of the blockchain as a whole. Ganache and Truffle are two of those tools that can be used to deploy contracts in a local blockchain network and simulate their functions in a responsive environment.

Next comes deploying contracts and testing them in the testnets, after the integration and testing work. More than anything, going to a testnet as Rinkeby, Ropsten, or Goerli will probably make the most sense as those simulate real-world conditions without the risk of using real tokens. Because these public testnets behave just like the Ethereum network, they make for a wonderful venue to test your project in a production-like setting. On the other hand, if you want to do a private blockchain, it is possible to set up a testnet using one of those frameworks like Hyperledger Fabric or Tendermint. Testing on these two will allow you to gauge how your blockchain project will perform while interacting using a broader community of nodes. In doing so, it would assist in dealing with some issues like network latency, transaction fees, and block propagation.

6 Result and Discussion

Conventional IP systems rely on established legal frameworks, such as patents, copyrights, and trademarks, which can be slow and costly. These systems often involve lengthy application processes, high legal fees, and challenges in proving ownership, making them less effective in the fast-paced digital landscape.

In contrast, blockchain platforms like Ethereum and Solana offer innovative solutions for IP management. Ethereum, with its robust smart contract capabilities, allows for the creation of decentralized applications that can automate IP rights management, enabling creators to register, track, and enforce their rights more efficiently. Its compatibility with various protocols enhances interoperability, making it a popular choice for IP-related projects.

Solana, known for its high throughput and low transaction costs, provides a scalable alternative for IP management. Its speed allows for real-time updates and transactions, which can be particularly beneficial for creators who need immediate validation of their rights. Solana's ecosystem supports various decentralized applications that can facilitate IP registration and licensing, making it an attractive option for innovators seeking efficient solutions. Both Ethereum and Solana address the challenges of conventional systems by providing transparency, security, and reduced reliance on intermediaries. They enable creators to maintain control over their IP assets while minimizing costs and delays associated with traditional methods. As the digital landscape continues to evolve, these blockchain-based solutions present a promising future for IP management, empowering creators and enhancing the protection of their IP.

The graph in Fig. 7 describes the cost efficiency of the blockchain tokens. It shows the drastic change and the growth of the types of blockchain that are available in the market.

It is popularly known that Solana stands out as the fastest blockchain capable of processing as many as 65,000 transactions per second (TPS). Nothing like this can compare to pre-Ethereum 2.0 at 30 TPS and less than 7 TPS on Bitcoin. Therefore, Solana provides the best platform for applications such as DeFi, gaming, and NFTs, which prioritize high-speed transactions. It is made possible through the combination of PoH and PoS consensus mechanisms. The best part is that with Proof of History (PoH), timestamps will be embedded in such blocks, which would replace the communication needed between nodes to establish the order, thus taking away the time it takes to confirm a transaction. This drastically improves throughput while maintaining low costs.

Compared to this is Ethereum, the second-largest blockchain by market capitalization. Yet, it suffers from issues of scalability. Though the planned switch of Ethereum into Ethereum 2.0 and the changeover from PoW to PoS promise better scalability, they still cannot touch the current transaction throughput of Solana. Ethereum faces very high gas fees and slow transaction times, especially during times of extreme network activity. Many of these problems are expected to be addressed in Ethereum 2.0, especially with the coming of sharding, but Solana is very much already ahead on greatly leading infrastructure runs in terms of transaction speed and efficiency. The comparison of various contracts is shown in Table 1.

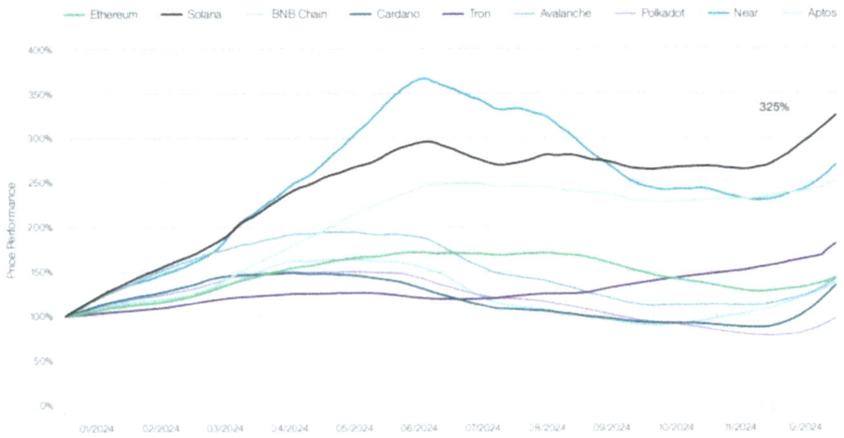

Fig. 7 Cost comparison

Table 1 Comparison of Ethereum versus Solana with various aspects

Aspect	Conventional system	Ethereum	Solana
Legal framework	Relies on established legal frameworks (patents, copyrights, trademarks)	Utilizes smart contracts for decentralized IP management	Employs smart contracts for efficient IP registration and licensing
Processing time	Lengthy application processes, often taking months or years	Faster due to automation, but still subject to network congestion	Very fast, enabling real-time updates and transactions
Cost	High legal fees, application costs, and expert consultations	Lower costs due to reduced need for intermediaries, but gas fees can vary	Low transaction costs make it affordable for creators
Ownership verification	Difficult to establish origin or ownership, leading to disputes	Transparent ownership records on the blockchain, easier to prove rights	Similar transparency, with quick updates to ownership records
Intermediaries	Requires lawyers and administrative bodies, increasing complexity and costs	Reduces reliance on intermediaries through automated processes	Minimizes intermediaries, streamlining the IP management process
Security	Vulnerable to infringement and exploitation due to slow processes	Enhanced security through blockchain technology, making IP harder to infringe	High security with fast transaction validation, reducing risks of fraud
Innovation support	Slower to adapt to new technologies and methods	Supports a wide range of decentralized applications (dApps) for IP management	Scalable platform for various dApps, facilitating innovative IP solutions
Transparency	Limited transparency in ownership and rights management	Full transparency of transactions and ownership on the blockchain	Transparent ecosystem with clear visibility of IP rights and transactions
User control	Creators often lose control over their IP due to complex legal processes	Creators maintain control over their IP assets through smart contracts	Empowering creators with direct control over their IP rights and assets
Future potential	Struggling to keep pace with rapid innovation and information sharing	Promising future with ongoing developments in blockchain technology for IP management	High potential for growth and adoption in IP management, leveraging speed and efficiency

Despite all commendations and potential of blockchain technology in IP preservation, the technology is faced with issues such as scalability, the energy consumption of certain consensus mechanisms, and difficulties in integrating blockchain with existing legal frameworks. Another barrier to the widespread use of blockchain IP-preserving systems is that industries may not easily let go of the ample resistance they have to change from traditional methods.

7 Conclusion

The IP project marks a huge leap forward—the revolutionary paradigm in the management of IP. The project is based on the application of blockchain technology that turns into an efficient, secure, and cost-effective platform. By allowing users to manage their own IP assets comprehension through a platform that diminishes the consumption of traditional intermediaries like lawyers and registries, both costly and time-consuming. This platform gains strength from emphasizing transparency and user control, which contributes to trust as well as streamlining IP registration and transfer processes. It is therefore possible for creators, inventors, and businesses to directly interact with the platform.

There are definitely challenges—a major one being educating users on blockchain technology along with navigating dynamic regulation. However, the advantages of this platform far outweigh these constraints. Education to users on ongoing needs in understanding blockchain technology is one such effort geared toward making the technology more friendly and comprehensible for use by many. Again, the innovation can transition and comply with new laws for IP and blockchain regime evolution while it advances. With the profile of growth of IP throughout changes of feedback from its user base and adaptable adherence to changing regulations, the project is headed toward revolutionizing IP management. Long-term benefits of this platform for creators, businesses, and innovators will be enormous in a new age of more open, efficient silos-neither the centralization nor the commanded trust environment will spread.

References

1. Fennell LA (2013) The problem of resource access. Harv Law Rev 1471–1531
2. Rana SK, Rana AK, Rana SK, Sharma V, Lilhore UK, Khalaf OI, Galletta A (2023) Decentralized model to protect digital evidence via smart contracts using layer 2 polygon blockchain. IEEE Access 11:83289–83300
3. Qizi LRM, Kamalovich BK (2025) Blockchain technology usage on intellectual property rights. Int J Semiot Law-Revue internationale de Sémiotique juridique 38(2):363–380
4. Alqarni A (2024) A blockchain-based solution for transparent intellectual property rights management: smart contracts as enablers. Kybernetes
5. Mazzarol T, Reboud S, Mazzarol T, Reboud S (2020) Intellectual property management. In: Entrepreneurship and innovation: theory, practice and context, pp 431–470

6. Kumar P (2024) Intellectual property rights (IPR): nurturing creativity, fostering innovation. Ideal J Adv Res Prog Spectr (IJARPS) 3(02):32–38. eISSN: 2583-6986
7. Bonnet S, Teuteberg F (2023) Impact of blockchain and distributed ledger technology for the management, protection, enforcement and monetization of intellectual property: a systematic literature review. Inf Syst e-Bus Manag 21(2):229–275
8. Bi Y (2025) Vibrating outside borders studying the universal outreach of chinese music arts through music interaction. Cultura: Int J Philos Cult Axiol 22(4)
9. Mary T, Enoch O (2024) Legal considerations in the development and commercialization of corporate intellectual property. Int J Rural Dev, Environ Health Res 8(3):01–20
10. Põld L (2023) Beyond traditional intellectual property: the necessity and possibilities of strengthening the protection of trade secrets through its integration into the modern intellectual property system. Int Comp Jurisprud 9(1):123–138
11. Lim D (2018) AI & IP: innovation and creativity in an age of accelerated change. Akron Law Rev 52:813
12. Carlson JJ, Walton SM, Basu A, Chapman RH, Campbell JD, McQueen RB, Pearson SD et al (2019) Achieving appropriate model transparency: challenges and potential solutions for making value-based decisions in the United States. Pharmacoeconomics 37:1321–1327
13. Berentsen A (2019) Aleksander berentsen recommends "bitcoin: a peer-to-peer electronic cash system" by Satoshi Nakamoto. In: 21st century economics: economic ideas you should read and remember, pp 7–8
14. Zyskind G, Nathan O (2015) Decentralizing privacy: using blockchain to protect personal data. In: 2015 IEEE security and privacy workshops. IEEE, pp 180–184
15. Kaur S, Jaswal N, Singh H (2022) Blockchain technology: principles, applications, and advantages of blockchain technology in the digital era. In: Applications, challenges, and opportunities of blockchain technology in banking and insurance. IGI Global, pp 204–212
16. Barman P (2024) The impact of blockchain technology on intellectual property (IP) in reshaping the secure digital environment. In: Libr Prog-Libr Sci, Inf Technol Comput 44(3)
17. Yang W, Shi ZJ, Lin S (2023) Art, smart contracts, and the role of royalties. In: HKUST Business School Research Paper 2023-109
18. Aveni A, Ulisse C (2025) Intellectual property and the future of NFT market. Revista Processus de Estudos de Gestão, Jurídicos e Financeiros 16(50):e501342–e501342
19. Bogdanov A, Khvatov V, Uteshev A, Shchegoleva N (2024) Virtual blockchain network for secure financial and industrial applications. In: International conference on computational science and its applications. Springer Nature Switzerland, Cham, pp 142–161
20. Buterin V (2024) Ethereum white paper: a next generation smart contract and decentralized application platform. First version 53
21. Mougayar W (2016) The business blockchain: promise, practice, and application of the next Internet technology. John Wiley & Sons
22. Alketbi S, Mahmuddin MB (2024) Intellectual property blockchain. In: 2024 6th international conference on blockchain computing and applications (BCCA). IEEE, pp 372–376
23. Zohar A (2015) Bitcoin: under the hood communications of the ACM, pp 104–113
24. Khan SN, Loukil F, Ghedira-Guegan C, Benkhelifa E, Bani-Hani A (2021) Blockchain smart contracts: applications, challenges, and future trends. Peer-to-Peer Netw Appl 14:2901–2925
25. Krajewski T, Lettiere R (2019) Efforts integrating blockchain with intellectual property. Les Nouvelles-J Licens Exec Soc 54(1)
26. Mentsiev AU, Guzueva ER, Yunaeva SM, Engel MV, Abubakarov MV (2019) Blockchain as a technology for the transition to a new digital economy. J Phys: Conf Ser 1399(3):033113
27. El Koshiry A, Eliwa E, El-Hafeez TA, Shams MY (2023) Unlocking the power of blockchain in education: an overview of innovations and outcomes. Blockchain: Res Appl 4(4):100165
28. Zubaydi HD, Varga P, Molnár S (2023) Leveraging blockchain technology for ensuring security and privacy aspects in internet of things: a systematic literature review. Sensors 23(2):788
29. Radziwill N (2018) Blockchain revolution: how the technology behind bitcoin is changing money, business, and the world, 2016. Dan Tapscott and Alex Tapscott. Penguin Random House, New York, pp 64–65, 348 pages

30. Antal C, Cioara T, Anghel I, Antal M, Salomie I (2021) Distributed ledger technology review and decentralized applications development guidelines. Futur Internet 13(3):62
31. Kumar KS, Ananth Kumar T, Radhamani AS, Sundaresan S (2020) Blockchain technology: an insight into architecture, use cases, and its application with industrial IoT and big data. In: Blockchain technology. CRC Press, pp 23–42
32. Wanigasooriya CS, Gunasekara ADAI, Kottegoda KGKG (2022) Blockchain-based intellectual property management using smart contracts. In: 2022 3rd international conference for emerging technology (INCET). IEEE, pp 1–5

Secure and Reliable Healthcare Credential Verification Using Blockchain

S. Madumidha, P. Sivaranjani, M. Nirosh Gowda, P. S. Selva Kirubha, K. Thamarai, K. Suresh Kumar, and Matthew Olusegun Adigun

Abstract An online platform that will revolutionize the management and verification of healthcare professionals' credentials has been developed. The system is developed on the principles of security, efficiency, and user-friendly access. The platform is a product built in Java on a blockchain-based backend system. This ensures maximum protection of data integrity and privacy through advanced encryption algorithms. A well-functioning credential management system is imperative for healthcare organizations as they hold sensitive information and manage risks, including data breaches, identity theft, and credential fraud. The system creates a safe environment to handle personal information pertaining to certificates and qualifications while ensuring strict privacy. The decentralized peer endorsement system is one of the key distinguishing features of the system. Unlike centralized systems, wherein credentials are vetted and certified through peer endorsements throughout the spectrum of trust and accountability, this system discourages fraudulent entries. Furthermore, a two-step verification system further strengthens the security of certifications authenticated from trusted authority providers such as medical boards or regulatory bodies, ensuring that validation of recorded information is done comprehensively and enhances the integrity of the credential verification system.

Keywords Authentication system · Peer endorsement algorithm · Blockchain technology · Data security · Secure credential management

---

S. Madumidha · M. N. Gowda · P. S. S. Kirubha · K. Thamarai
Department of Information Technology, Sri Krishna College of Technology, Coimbatore, Tamil Nadu, India
e-mail: madumidha.s@skct.edu.in

P. Sivaranjani
Department of ECE, Kongu Engineering College, Erode, India

K. S. Kumar (✉)
Department of Computer Science and Engineering, Sri Eshwar College of Engineering, Coimbatore, India
e-mail: sureshkumarkskphd@gmail.com

M. O. Adigun
Department of Computer Science, University of Zululand, Kwadlangezwa, South Africa

A. Kumar et al. (eds.), *Blockchain and Federated Learning Synergy for Privacy-Focused DeepFex Solutions*, Blockchain Technologies,
https://doi.org/10.1007/978-981-95-1394-9_11

1 Introduction

Credentialing and verification are the most important parts of the healthcare industry, which concern those who authenticate and veterinarian professionals for patient care. Traditional credentialing often proves lengthy, complicated, and beset with problems like poor data security, difficulty in proving the authenticity of credentials, and the possibility of fraud [1]. Major implications of credential fraud are that unqualified individuals may take care of a sick patient, which means a patient's safety is at stake [2]. Secure file-sharing thus forms a security feature to make these efforts easier and more user-friendly. It creates a way of encrypting files like credential documents and password keys, which would then be transmitted through secure channels to privileged verifiers. The verifiers then have to decode the encrypted files with the issued keys, thus barred from gaining access to sensitive information [3]. Critical to this system is establishing an immutable, secure ledger with an auditable form of transparency; building this dimension into the system design is and will remain blockchain technology. Each activity performed within the platform—from submitting credentials for verification—is registered to the blockchain, creating a tamper-proof and trackable record of all actions [4]. This greatly reduces the risk of fraud but guarantees accountability and trust in the whole credentialing process.

This makes the system easier for the organizations to follow movement and compliance with regulatory standards. Also, our platform is scalable and user-friendly. Hence, it is suitable for any health organization, no matter what size, to adopt it without much technical overhead. The interface is so intuitive that users would not find it hard to learn how to navigate using the platform, thus reducing onboarding time toward encouraging wider adoption [5]. The multilayered security and authentication provided by the platform make it a whole solution to manage healthcare credentials, integrating the blocking of fraud using credentials through security by blockchain and encryption privacy with multilayered authentication. This boosts the trust and operational efficiency in the healthcare institutions because it not only protects against credential fraud but also provides a faster and more reliable way to verify credentials [6]. This is the rationale for having a platform that shows how technology can be employed for security enhancement and process streamlining in sensitive industries like healthcare.

2 Literature Survey

Increased attention has recently been accorded to the credentialing processes of health care practitioners. This has brought increased emphasis to the seriousness with which the healthcare organizations guarantee competencies and experience qualifications of credentialed healthcare providers before access is made to any facility [7]. Such a vital safety provision empowers institutions to evaluate their health professionals—doctors and dentists as well as allied health workers—and witnesses the verification of their backgrounds and documents such as educational tests, certifications, and licenses, with a view to patient safety, legal rights, and financial liabilities with respect to employing unqualified individuals [8, 9].

Machine-to-machine communications, wearable health devices, and virtual health services are going to play key roles in meeting increasing demands and relieving some workloads of professional caregivers by automating standard procedures and enabling nonprofessionals to perform a few health-related tasks [10]. In parallel, a newcomer in the health systems, the most promising emerging innovation, has turned out to be blockchain technology that could revolutionize healthcare by securing more accessibility in Electronic Health Record (EHR) systems [11].

Literature further suggests that with this decentralized and tamper-proof nature, blockchain can significantly improve the storage and sharing of sensitive medical data, thus becoming a very useful tool for health IT infrastructure [12]. These findings can serve as a guiding compass for researchers and health organizations in selecting the most suitable blockchain model for secure EHR management [13].

This decentralized structure of blockchain would be very useful for resolving the problem of sharing electronic medical records (EMRs), which are locked to various hospitals and lack interoperability [14]. It hardly allows the sharing of data within healthcare facilities, where data is being shared within the healthcare system. Features like automated provenance tracking and credible credential verification of data further boost the trustworthiness of the data-sharing process [15].

And besides, blockchains can be used to enhance security features, such as empowering patients through more control of their health data [16]. People can manage and share information with a healthcare institution in a secure manner, granting and revoking access as they want or need to—within such clear, accessible terms [17]. Thus, visibility paved the way toward a more collaborative atmosphere of letting doctors and patients see an evidence-based, real-time picture of that dynamic interface. Quite simply, applying blockchain technology would mean more active participation of the patient in his care, thereby improving health outcomes and satisfaction with the system [18].

Blockchain-based credentialing involves automated and trustless verification mechanisms through smart contracts. In the review related to smart contracts as a part of blockchain. It discussed the prospects of smart contracts in credential validation through automated execution [19]. Further highlighted smart contract use cases like automated licensing processes and peer-based verification mechanisms. Notwithstanding these merits, smart contracts should be legally aligned, warned [20], since immutably kept credential records may violate personal data protection laws such as the GDPR and HIPAA. This was also stated by [5], who added that there is a necessity for flexible regulatory frameworks that can address blockchain-based credentialing.

Turning to cases of interoperability, the authors in [21] proposed a framework for blockchain-healthcare systems, which envisaged secure and real-time credential verification across disparate platforms. The authors in [22] have also reviewed the applications of geospatial blockchains in healthcare, emphasizing their use for location-based access control in credential verification.

For the future, the research ought to optimize blockchain protocols for credential verification in terms of efficiency and scalability. The systematic review of blockchain applications in the healthcare sector found that scalability and compliance are two of the major hindrances to blockchain implementation. While the authors in [23, 24] discussed blockchain ecosystems for verifiable qualifications, they also highlighted the applicability of such ecosystems in educational, employment, and healthcare credentialing.

Another trail worth pursuing is zero-knowledge proofs (ZKPs), where verification of a credential does not mean exposure of the credential itself. They believed ZKPs could promote privacy compliance while preserving data integrity. Also, the authors [25, 26] have suggested deploying AI-driven fraud detection mechanisms with a view to enhancing credential verification accuracy.

3 Problem Statement

Nowadays, many hospitals and organizations continue to use old technologies or ad hoc semiautomated processes that are scattered among many intermediaries and multiple data sources [27]. The slow processes in verifying healthcare professionals, delaying their onboarding, increase the administrative backlog and create adverse effects on patient care and overall operations of the hospitals [28]. Beyond the inconvenient dates, there are long waits between these ends.

In order to solve these important problems, we have constructed a secure web-based platform that takes advantage of modern technologies such as Java and encryption algorithms in blockchain [29]. Our solution offers credential management that is decentralized, transparent, and ensures that once professional credentials have been entered into the system, they cannot be tampered with and changed without detection, as may be the case normally [30]. A feature unique to the process of using blockchain technology is the immutability of credentials; this increases the integrity of the credentialing process [31]. What sets our platform apart is its multi-layered authentication process that combines peer endorsement and validation by trusted authorities, ensuring that only an appropriate person will use and verify the credentials [32]. Additionally, these processes improve the integrity of the data while increasing accountability at the various levels [33]. Peer endorsement, plus trusted authorities validating the credentials, creates an environment of transparency and reliability in the healthcare sector.

3.1 Current Mechanism

The existing approach is very expensive and time consuming, often requiring a lot of days to process. The existing methods involve manual checks, third-party verification services, or reliance on paper-based documents, which are inefficient and vulnerable to manipulation [34]. There is no centralized, tamper-proof system that allows healthcare institutions and employers to quickly and reliably verify the certifications of healthcare professionals. Moreover, people are using fake documents to enhance their resumes to get jobs. Lack of security leads to a lot of fraudulent certificates being created in a minimum amount of time [35].

Figure 1 is a blockchain-based system for the verification of healthcare professionals' credentials. It consists of three primary stakeholders: healthcare institutions, credential issuers, and professionals. The healthcare institutions input information regarding professionals, which credential issuers authenticate and store on the blockchain safely. This blockchain record enables professionals to receive authenticated credentials, which they can present to employers or verifiers for employment application or compliance purposes. The system will provide safe, open, and tamper-proof verification of credentials.

First, medical workers apply to a health organization for verification of their credentials by sending all required documents. The documents are then sent to primary verifiers; that is, a verifier and an authority provider. These are the people who do the first verification—who check whether the claimed credentials are valid.

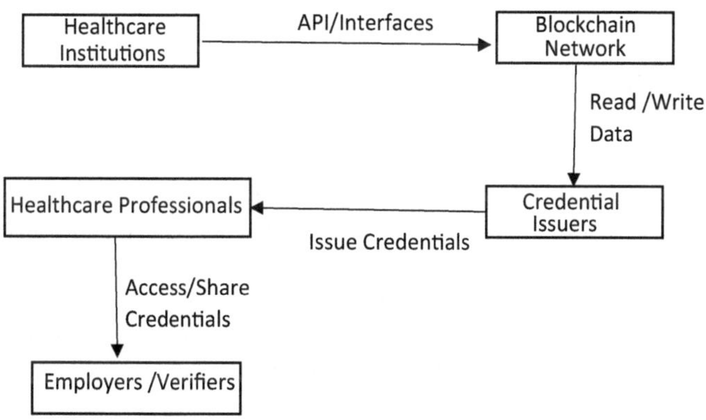

Fig. 1 Architecture diagram of the existing system

Table 1 Comparison of existing system and proposed system

Features	[1]	[10]	[12]	[24]	Our solution
Mode of operation	Private	Private	Consortium	NA	Private
Verification algorithm	NA	NA	NA	NA	Peer endorsement
Encryption mechanism	Keccak256	NA	NA	NA	AES encryption
Digital signature	Yes	Yes	Yes	No	Yes

Upon completion of primary verification, validated information is provided to the medical staff department for secondary checks. The staff performs a detailed analysis of the documents stated. If the credentials are up to the standards, they get approved; conversely, they are rejected. The final verdict of approval or disapproval is then communicated to the healthcare professional. Definitely, this system comes with major limitations. It is slow, primarily manual verification, leading to many delays and inefficiencies. There are also multiple intermediaries, enhancing the opportunity for error and fraud in the system.

On the contrary, the absence of centrality and a tamper-proof mechanism makes it impossible to prevent the abuse of fake credentials for fraudulent employment. Since blockchain enhances the effectiveness of verification against fraud, it would make the process more secure and transparent and reduce any reliance on manual checks. Table 1 defines the comparison between the existing system and proposed system features. But a significant hurdle is making medical professionals use this blockchain-based verification process. It also needs technical expertise to develop and maintain the system, which will discourage wider use.

4 Methodology

The proposed approach encrypts sensitive data using Advanced Encryption Standard (AES) techniques, converting it into an unreadable format by means of a secret key via authorized users' access control. The data owner first encrypts the information along with the predefined access policy, after which he uploads the encrypted information to a cloud server, which stores the encrypted information securely. A valid user is verified for identity and requested access; the cloud server checks and grants access, provided it is approved by both server and data owner.

Only a single decryption key is given to the individual, and using this, one can securely download and view the files. Users have to make their choice regarding files to ensure the integrity of the session; later, they are taken back to the homepage. Not only this, but there are multilayered security protocols in the cloud server which will be able to observe the user and every present activity around, hence able to check amazingly of any unauthorized access attempt. The system would also be compliant with many other standards and thus can be used in sectors that demand robust data protection. Another unique feature provided by our system is that it provides the data owner with complete control over the encryption keys, such that even the cloud service provider does not get access to the encrypted files without the owner's permission. This brings an added level of privacy, where very sensitive data stays confidential at all times.

4.1 Algorithms

Algorithm 1 explains the encryption process of the file uploaded by the user at the time of registration. It generates the secret key which is used to decrypt the encrypted file. Algorithm 2 explains the decryption process. In decryption, the encrypted information will be decoded and decrypted with a secret key before being converted back into text. Algorithm 3 illustrates the peer endorsement algorithm, which is used to verify the user under some defined conditions using blockchain. Algorithm 4 explains the Proof of Authority (PoA), where the blocks are created based on the round robin method by validating the transactions.

Algorithm 1 AES-Based File Encryption Process

Input: PlainTextCredential, SecreteKeyOutput: EncryptedCredential

BEGIN: Generate an AES key using a key generation function.

Initialize the AES Cipher in encryption mode using SecretKey.

Encrypt the PlainTextCredential using the AES Cipher.

Convert the encrypted output to a base64 string to store in the blockchain.

Return EncryptedCredential.

END

Algorithm 1 describes the AES-based file encryption process, which encrypts confidential data with the AES. It takes two inputs: PlainTextCredential (the credential to be encrypted) and SecretKey (a secure encryption key). It first creates an AES key using a key generation function in order to guarantee a unique and secure encryption process. The AES Cipher is subsequently initialized in encryption mode with the given SecretKey, ready for the encryption process. The algorithm goes ahead to encrypt the PlainTextCredential using the AES Cipher, transforming the readable data into an unreadable ciphertext. This encrypted data is subsequently encoded into a base64 string, which is easy to store and transmit, especially in blockchain systems. Lastly, the EncryptedCredential is returned by the algorithm, which is the safely encrypted version of the original information. This is how confidentiality and integrity of data are assured, since only authorized users possessing the right SecretKey can decrypt and view the original data.

Algorithm 2 AES-Based File Decryption Process

Input: EncryptedCredential, SecretKeyOutput: DecryptedCredential

BEGIN: Decode the base64 encoded EncryptedCredential.

Initialize the AES Cipher in decryption mode using SecretKey.

Decrypt the EncryptedCredential using the AES Cipher.

Convert the decrypted output to a string format.

Return DecryptedCredential.

END

Algorithm 2 describes the AES-Based File Decryption Process, which is employed to decrypt encrypted data back to its original state. It takes two inputs: EncryptedCredential (encrypted data) and SecretKey (secure key for decryption). The algorithm first decodes the base64 encoded EncryptedCredential to get the original encrypted data in binary format. Subsequently, it sets up the AES Cipher to decrypt mode from the given SecretKey, in anticipation of reversing the encrypting process. The algorithm proceeds to decrypt the EncryptedCredential with the AES Cipher, translating the ciphertext back into its original readable form. This decrypted information is then translated further to string format to make it readable for humans. The algorithm then returns the DecryptedCredential, which is the initial plaintext data. This process guarantees only the correct individuals with the correct SecretKey may view the actual data, retaining confidentiality and safety.

```
Algorithm 3 Peer Endorsement Integrity for Sign up
Input: UserDetails, PeerEndorsements, BlockchainHash
BEGIN
    if UserDetails are incomplete or invalid
    then
    Return "Sign-Up Failed: Invalid User Details"
    Initialize EndorsementHashList as an empty list
    for each endorsement in PeerEndorsements do
  Compute Hash = HashFunction (endorsement)
  Append Hash to EndorsementHashList
Compute AggregatedHash = AggregateHashFunction (EndorsementHashList)
if AggregatedHash == BlockchainHash   then
     Store UserDetails and PeerEndorsements in the database
Return "Sign-Up Successful"
    Else Return "Sign-Up Failed: Endorsement Integrity Verification Failed"
     END
```

Algorithm 3 describes the peer endorsement integrity process for signing up users, validating user endorsements through blockchain verification. It has three inputs: UserDetails (user-provided information), PeerEndorsements (endorsements by trusted peers), and BlockchainHash (a reference hash kept on the blockchain for verification).

The algorithm first validates if the UserDetails are valid and complete. If they are invalid or incomplete, it returns "Sign-Up Failed: Invalid User Details" directly. If the UserDetails are valid, it creates an empty EndorsementHashList to hold hashed peer endorsements. It then iterates over the PeerEndorsements list, calculates a Hash using a secure HashFunction for each endorsement, and adds it to the Endorsement Hash List.

After hashing all endorsements, the algorithm computes an AggregatedHash by aggregating the individual hashes with an AggregateHashFunction. The algorithm then checks the AggregatedHash against the given BlockchainHash. If they are equal, the algorithm stores the UserDetails and PeerEndorsements securely in the database and returns "Sign-Up Successful.“ If they are not equal, the algorithm returns "Sign-Up Failed: Endorsement Integrity Verification Failed," which means there might be tampering or invalid endorsements. The process guarantees the credibility and legitimacy of user testimonials with blockchain verification.

```
Algorithm 4 Proof of Authority Consensus
BEGIN
   Initialize validators = [Validator1, Validator2, Validator3]
      Set roundRobinOrder = [Validator1, Validator2, Validator3]
   Set currentValidatorIndex = 0
   WHILE True:
  currentValidator=roundRobinOrder[currentValidatorIndex]
// Block Proposal
IF currentValidator is active:
        pendingTransactions = collectPendingTransactions ()
   newBlock = createBlock(pendingTransactions)
         signBlock (newBlock, currentValidator.privateKey)
broadcastBlock(newBlock)
// Block Validation
FOR each validator in validators
    IF validator receives newBlock:
         IF verifyBlockSignature (newBlock, currentValidator.publicKey) AND
validateTransactions(newBlock):
                    signBlock (newBlock, validator.privateKey)
               broadcastApproval(newBlock):
 // Consensus
      IF countApprovals(newBlock) >= majorityThreshold:
         finalizeBlock(newBlock)
         addToBlockchain(newBlock)
         updateLedgerState(newBlock)
         currentValidatorIndex = (currentValidatorIndex + 1) %
length(roundRobinOrder)
 // Fault Tolerance
      IF currentValidator fails to propose block within timeout: currentValidatorIndex
= (currentValidatorIndex + 1) % length(roundRobinOrder)
 END
```

10

Algorithm 4 describes the PoA Consensus algorithm, which is based on a set of preapproved validators to verify and append blocks to the blockchain. This model of consensus provides fast and secure block verification using a round-robin approach. The algorithm begins by initializing a list of reliable validators (Validator1, Validator2, and Validator3) and setting roundRobin Order to decide the order in which validators would propose new blocks. It also initializes the current Validator Index to keep track of which validator would propose the next block. In an infinite loop, the algorithm chooses the current Validator according to the roundRobin Order. If the chosen validator is active, they gather pending transactions, create a new block with these transactions, sign the block with their private key, and broadcast it to other validators.

When other validators receive the new block, they check its signature with the private key of the proposer and certify the transactions contained within it. When the block is legitimate, every validator signs the block with their private key and broadcasts their endorsement. For a block to get added to the blockchain, it has to have approval from the majority of the validators. After gathering the approvals needed, the block gets finalized, gets added to the blockchain, and the ledger state gets updated. The algorithm updates the current Validator Index to choose the next validator in the round-robin list. If the current Validator does not suggest a block within a given timeout, the algorithm automatically switches to the subsequent validator in the list. This is a fault-tolerance feature that allows block generation to proceed uninterrupted despite a single validator going offline. This model of PoA delivers efficient consensus with minimum computational overhead while ensuring security through responsibility of validators.

5 Designed System

The designed system is a secure web platform which will provide supportive healthcare management professionals for credential holding by Java blockchain and enhanced encryption methods. Security provisions for identity protection will include making access requests anonymous—both cloud server authentication and data owner approval will be required before giving access to any data, thus ensuring an extra level of security.

Figure 2 illustrates the existing system for verifying healthcare professionals' credentials through a multistep process. It begins when healthcare professionals apply to a healthcare organization by submitting all the necessary documents. These applications are then forwarded to trusted primary verifiers, which include a verifier and an authority provider, responsible for the initial validation of the submitted information. After the primary verification, the validated data is sent to the medical staff department, where further analysis and verification are performed on the uploaded files. This step ensures a detailed examination of the credentials and their authenticity. The status of the documents—whether approved or rejected—is updated and sent back to the healthcare professionals, informing them of the outcome. This system involves multiple layers of verification to ensure the accuracy and reliability of professional credentials. However, it relies on manual processes and various intermediaries, which can lead to delays and inefficiencies. Implementing a blockchain-based system

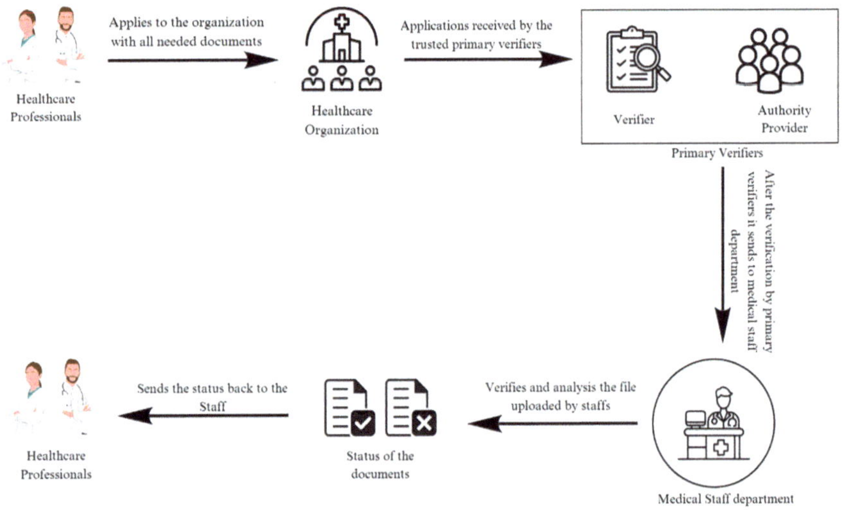

Fig. 2 Process of the system

could improve transparency, security, and efficiency by reducing manual checks and providing a tamper-proof record of verified credentials.

The proposed system is developing a blockchain and smart contract-based solution for automating and streamlining the process of credential verification, as one of those challenges. A decentralized, conspicuously designed tamper-proof ledger serves as the foundational infrastructure on which basic acts of immutability and very straightforwardness will be supposed to occur in records. Automated verification tasks are applied by smart contracts to tackle the problem. When healthcare professionals submit their credential information, a smart contract is triggered to validate them against predefined criteria and does not require primary verifiers or the medical staff department to check data manually. Once verified, the smart contract updates the blockchain with the verification status, creating an immutable and transparent record, accessible by authorized parties only. Thus, all intermediaries are avoided; however, this accelerates the time consumed, dramatically improving the efficiency. In addition, the enhanced encryption methods and dual-layer authentication mechanism used by the system guarantee a high level of security and privacy for sensitive data. Combining blockchain, smart contracts, and advanced encryption constructs a more transparent integration for credential verification, making it more secure and efficient compared with existing manual processes and providing a credible environment for healthcare management professionals.

6 Results and Discussion

The intended purpose of this secure web-based platform for managing and verifying credentials of health care professionals is such that it supersedes any traditional system in doing so by virtue of its use of blockchain technology and sophisticated encryption methods in managing and securing credentials. This will almost eliminate all previous shortcomings of credential management systems and will create much greater integrity, security, and efficiency for all data produced in the platform. The use of blockchain as an emerging technology forms a tamper-proof source of any transaction or record of credentials, making it almost impossible for people to alter or manipulate the data unauthorizedly. It is a huge step forward from the conventional centralized databases, where tampering and fraud are common.

Figure 3 shows the system user registration page where new medical staff and health professionals can key in their minimal information for crediting verification purposes. Users will have to include key information, including name, e-mail, password, address, phone number, field of interest, board certification, name of certification course, and work history. The system includes the registration of users within the peer endorsement integrity algorithm that verifies credentials among the health workers electronically. This automated process greatly minimizes administrative burden and speeds up verification, allowing for quicker registration and enhanced efficiency. Moreover, the system has a double security feature in which all credential information needs to be validated by trusted authority providers. This additional layer of authentication guarantees that only legitimate and verified data is saved on the platform. The integration of peer recommendations and authority validation increases data integrity and security, reducing the possibility of fake entries. This design of the system not only guarantees quicker credential verification but also ensures transparency and safeguards sensitive data during the registration process.

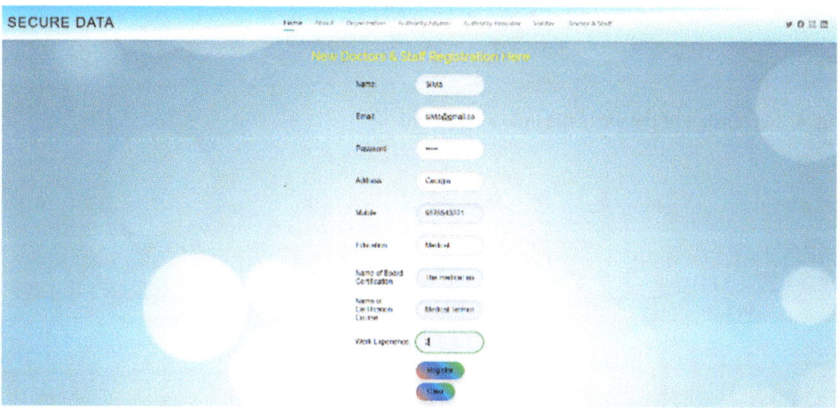

Fig. 3 Screenshot of user registration page

Figure 4 illustrates the system's file details page, displaying a table containing details of files uploaded. The table has columns for File ID, User ID, Owner Name, Mail, File Name, Upload Date, and Remove. Each row in the table refers to a unique file uploaded by a user with information on the file name, email of uploader, and upload date. The "Delete" link in the remove column can be used to delete files. This section is useful for managing and monitoring uploaded credentials, making it easy to access and administratively control them. The addition of file ownership information makes accountability and data traceability easier in the system.

Figure 5 represents the example of an email containing a One-Time Password (OTP) sent to a user or to a verifier. An OTP is utilized as a security feature to authenticate that a particular user is entitled to download certain documents. The illustrated figure explains the real-life setup of how an OTP works, displaying the format of the email and the manner of receiving the OTP. The importance of this OTP is in providing secure access for only intended users to download the document. This is a crucial step to ensure data security and prevent unauthorized access.

Table 1 provides a comparison of the new healthcare credential security system with current systems on the basis of important features like mode of operation, verification algorithm, encryption mechanism, and digital signature. The new system runs within a private environment like most current solutions for ensuring controlled

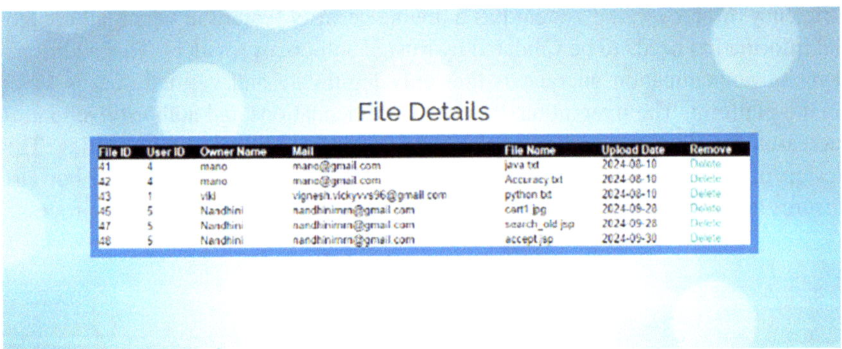
File Details

File ID	User ID	Owner Name	Mail	File Name	Upload Date	Remove
41	4	mano	mano@gmail.com	java.txt	2024-08-10	Delete
42	4	mano	mano@gmail.com	Accuracy.txt	2024-08-10	Delete
43	1	viki	vignesh.vickyvvs96@gmail.com	python.txt	2024-08-10	Delete
45	5	Nandhini	nandhinimm@gmail.com	cart1.jpg	2024-09-28	Delete
47	5	Nandhini	nandhinimm@gmail.com	search_old.jsp	2024-09-28	Delete
48	5	Nandhini	nandhinimm@gmail.com	accept.jsp	2024-09-30	Delete

Fig. 4 Screenshot of the users file uploaded record

Fig. 5 Screenshot of getting e-mail for OTP

access as well as increased data privacy. The private blockchain platform provides a secure and trusted approach to handling sensitive healthcare credentials while ensuring confidentiality. One of the major improvements in the suggested system is the implementation of peer endorsement as a verification algorithm. Other systems that do not have automated verification processes (NA) cannot compare with peer endorsement, which enables healthcare workers to authenticate one another's credentials quickly, minimizing manual intervention and administrative delays. The suggested system also uses AES encryption, a strong and common encryption standard, to protect sensitive information. Conversely, just a single one among the compared solutions utilizes Keccak256, with the rest using no encryption at all (NA), thus having higher vulnerability to data breaches. The system is also equipped with digital signatures like the majority of available solutions for the purpose of authenticating and maintaining the integrity of credential data. With all these advanced aspects, the developed system provides more enhanced security and efficiency than traditional solutions, supporting higher reliability for healthcare credential validation.

6.1 Discussion

The proposed online platform for healthcare credential management and verification constitutes a truly transformative change from old-fashioned, manual methods of verification to one that is a secure, blockchain-based one. The platform has a strong inclination toward secure, efficient, and easy access so that healthcare professionals can have their credentials managed with utmost integrity and reliability.

One unique feature of this system is the application of blockchain technology on the backend, providing a decentralized and tamper-proof ledger for credential-type data storage. Such an arrangement will ensure that sensitive healthcare data is immutable and immune to any form of fraudulent modification. The use of state-of-the-art encryption techniques such as AES and peer-endorsing algorithms will prevent unauthorized access, identity theft, and data breaches, thus making their contributions to the system's degree of acceptance. In addition, by preventing reliance on a third-party centre for the storage of these records, this system of management reduces, to a great degree, the chances associated with the risks of single points of failure, thus fostering the overall strength of the system.

The most important phase in system development life cycle is system testing. The number and nature of errors in a newly designed system depend on the system specifications and the time frame given for the design. A newly designed system should have all the subsystems working together, but in reality, each subsystems work independently. During this phase, all the subsystems are gathered into one pool and tested to determine whether they meet the user requirements.

Testing is done at two levels: Testing of individual modules and testing the entire system. During the system testing, the system is used experimentally to ensure that the software will run according to the specifications and in the way the user expects. Each test case is designed with the intent of finding errors in the way the system will process it. Testing plays a very critical role in determining the reliability and efficiency of software and hence is a very important stage in software development. Software testing is done at different levels. They are the unit testing and system testing, which comprises integration testing and acceptance testing.

System implementation is the stage in the project where the theoretical design is turned into a working system. The most critical stage is achieving a successful system and in giving confidence on the new system for the user that it will work efficiently and effectively. The existing system was a long process. The project execution was checked with live environment, and the user requirements are satisfied. Proper implementation is essential to provide a reliable system to meet the organization's requirements.

6.2 Advantages

- The integration of encryption algorithms and blockchain technology creates a robust security framework, effectively protecting sensitive healthcare credentials from unauthorized access.
- Blockchain ensures that all credential data is tamper-proof and transparent, enhancing trust in the authenticity of healthcare professionals' qualifications.
- Automated peer endorsements streamline credential verification, significantly reducing administrative delays and speeding up the hiring process.
- Implementing double security measures by authority providers adds an extra layer of protection for credentials and sensitive files, ensuring authorized access only.
- Sending password keys via email ensures that only verified users can access and download sensitive files, minimizing the risk of data breaches.

6.3 Limitations and Challenges

- It is actually very expensive to store large files on blockchain. They are required to have off-chain storage solutions such as IPFS or cloud storage as an alternate solution.
- Most hospital activities and health care organizations are linked to legacy credential management systems, which may not support integration into blockchain.
- An investment in servers, storage, and skilled personnel is expensive to deploy and maintain blockchain infrastructure. Smaller health care will find it not cost-effective.
- Healthcare professionals may be hesitant to adopt it due to the lack of familiarity and mistrust of new technologies. Otherwise, awareness and training programs are advocated for smooth adoption.

7 Conclusion

This chapter offers a secure web platform for managing and verifying credentials of healthcare professionals. Employing blockchain and encryption technologies, it surpasses traditional system redundancy and security risks by ensuring the fact that credential data stays in possession of the institutions, while being tamper-proof and transparent. Some features like automated peer endorsements and double-layered security improve trust and reduce opportunities for fraud while expediting the verification process. Thus, paving the way toward further secure and reliable credential management processes.

Future developments could include integrating artificial intelligence and machine learning to enhance credential verification and detect anomalies. Expanding the platform to offer continuous monitoring of licenses and certifications would ensure healthcare professionals maintain their qualifications. Increasing interoperability with other healthcare systems could facilitate seamless data sharing, further enhancing efficiency. Addressing the limitations of blockchain adoption such as costs and technical knowledge through educational initiatives can promote wider implementation. Overall, the future scope encompasses technological advancements, expanded functionalities, and strategic partnerships to revolutionize healthcare credential management.

References

1. Alnuaimi A, Hawashin D, Jayaraman R (2020) Trustworthy healthcare professional credential verification using blockchain technology. J Healthc Inf Secur 12(4):78–85
2. Hasselgren A, Rensaa J (2021) Blockchain for increased trust in virtual healthcare: a proof-of-concept study. Int J Healthc Manag 14(2):102–110
3. Bittins S, Kober G, Margheri A, Masi M (2021) Healthcare data management using blockchain technology. Health Data Sci J 6(1):67–75
4. Tyagi AK, Seranmadevi R (2022) Blockchain for enhancing security and privacy in smart healthcare. J Cybersecur Healthc 9(1):88–95
5. Hölbl M, Kompara M, Kamišalić A, Nemec Zlatolas L (2018) A systematic review of the use of blockchain in healthcare. Symmetry 10(10):470
6. McGhin T, Choo KKR, Liu CZ, He D (2019) Blockchain in healthcare applications: research challenges and opportunities. J Netw Comput Appl 135:62–75
7. Skiba DJ (2017) The potential of blockchain in education and health care. Nurs Educ Perspect 38(4):220–221
8. Mamun Q (2022) Blockchain technology in the future of healthcare. Smart Health 23:100223
9. Thenmozhi M, Dhanalakshmi R, Geetha S, Valli R (2021) Implementing blockchain technologies for health insurance claim processing in hospitals. Mater Today: Proc
10. Arenas R, Fernandez P (2018) CredenceLedger: a permissioned blockchain for verifiable academic credentials. In: Proceedings of international conference on engineering, technology and innovation (ICE/ITMC), pp 1–6
11. Shakor MY, Khaleel MI, Safran M, Alfarhood S, Zhu M (2024) Dynamic AES encryption and blockchain key management: a novel solution for cloud data security. IEEE Access 12:26334–26343

12. Bokariya PP, Motwani D (2021) Decentralization of credential verification system using blockchain. Int J Innov Technol Explor Eng 10(11):113–117
13. Kolavennu SR, Reddy GR, Shreya L, Vishal K, Shubha N (2024) Enhancing cloud data security: dynamic AES encryption with blockchain key management. Int J Eng Res Dev 20(5):145–149
14. Andrew J, Isravel DP, Sagayam KM, Bhushan B, Sei Y, Eunice J (2023) Blockchain for healthcare systems: architecture, security challenges, trends, and future directions. J Netw Comput Appl 236:103633
15. Chinedu-Eneh E, Ramaswamy P, Farmer PE, Farmer P (2024) Blockchain-based healthcare credentialing: a solution to high costs and administrative burdens. Cureus 16(6)
16. Khezr S, Moniruzzaman M, Yassine A, Benlamri R (2019) Blockchain technology in healthcare: a comprehensive review and directions for future research. Appl Sci 9(9):1736
17. Taherdoost H (2023) Smart contracts in blockchain technology: a critical review. Information 14(2):117
18. Mohanta BK, Panda SS, Jena D (2018) An overview of smart contract and use cases in blockchain technology. In: 2018 9th international conference on computing, communication and networking technologies (ICCCNT). IEEE, pp 1–4
19. Morales-Alarcón CH, Bodero-Poveda E, Villa-Yánez HM, Buñay-Guisñan PA (2024) Blockchain and its application in the peer review of scientific works: a systematic review. Publications 12(4):40
20. Namasudra S, Sharma P, Crespo RG, Shanmuganathan V (2023) Blockchain-based medical certificate generation and verification for IoT-based healthcare systems. IEEE Consum Electron Mag 12(2):83–93
21. Zarour M, Ansari MTJ, Alenezi M (2021) Evaluating the impact of blockchain models for secure and trustworthy electronic healthcare records. J Health Inform 15(3):45–52
22. Kamel Boulos MN, Wilson JT, Clauson KA (2018) Geospatial blockchain: Promises, challenges, and scenarios in health and healthcare. Int J Health Geogr 17(1):25
23. Kumar KS, Radha Mani AS, Sundaresan S, Ananth Kumar T, Harold Robinson Y (2021) Blockchain-based energy-efficient smart green city in IoT environments In: Blockchain for smart cities. Elsevier, pp 81–103
24. Garg R (2021) Blockchain ecosystem for education and employment verification. In: 13th international conference on network and communication security, Toronto Canada
25. Raghuwanshi P (2024) AI-driven identity and financial fraud detection for national security. J Artif Intell Gen Sci (JAIGS) 7(01):38–51. ISSN: 3006-4023
26. Johora FT, Hasan R, Farabi SF, Alam MZ, Sarkar MDI, Al Mahmud MdA (2024) AI advances: enhancing banking security with fraud detection. In: 2024 First international conference on technological innovations and advance computing (TIACOMP). IEEE, pp 289–294
27. Rani U, Pesole A, Vázquez IG (2024) Algorithmic management practices in regular workplaces: case studies in logistics and healthcare. Publications Office of the European Union, Luxembourg
28. Franco PM, Adams LR, Booth SC (2024) From onboarding onwards. In: Contemporary topics in patient safety, vol 3, p 9
29. Aliya B, Olga U, Yenlik B, Sogukpinar I (2023) Ensuring information security of web resources based on blockchain technologies. Int J Adv Comput Sci Appl 14(6)
30. Mazzocca C, Acar A, Uluagac S, Montanari R, Bellavista P, Conti M (2025) A survey on decentralized identifiers and verifiable credentials. IEEE Commun Surv Tutor
31. Ayub Khan A, Laghari AA, Shaikh AA, Bourouis S, Mamlouk AM, Alshazly H (2021) Educational blockchain: a secure degree attestation and verification traceability architecture for higher education commission. Appl Sci 11(22):10917
32. Alsheavi AN, Hawbani A, Othman W, Wang X, Qaid G, Zhao L, Al-Dubai A et al (2025) IoT authentication protocols: challenges, and comparative analysis. ACM Comput Surv 57(5):1–43

33. Adepoju AH, Austin-Gabriel B, Eweje A, Hamza O (2023) A data governance framework for high-impact programs: reducing redundancy and enhancing data quality at scale. Int J Multidiscip Res Growth Eval 4(6):1141–1154
34. Basir MdS, Buckmaster D, Raturi A, Zhang Y (2024) From pen and paper to digital precision: a comprehensive review of on-farm recordkeeping. Precis Agric 25(5):2643–2682
35. Lebopa LV, Muchenje T, Mathonsi TE, Maswikaneng SP (2024) Information verification system to prevent academic fraud by employees. In: 2024 International conference on artificial intelligence, big data, computing and data communication systems (icABCD). IEEE, pp 1–6

Leveraging Blockchain Technology to Combat Deepfakes: A Novel Approach

Ramesh Yegireddi, Gudla Jayanth Sairam, Vandrangi Sandeep, Mailarabhatla Sirisha, B. L. Sivamani, and Moyyi Premsai

Abstract In today's highly connected world, there are even greater risks facing online media content for security. We have approached this challenge by developing a new security system that brings together three strong technologies: Blockchain for open ledger, InterPlanetary File System (IPFS) for distributed storage, and AI-assisted deep learning for smart verification. We developed it using the VGG16 neural network model on a web platform so it would be efficient and easy to use. We start by completely stripping identifying metadata and deriving important features from the media files, and safely encrypting them by using AES technology. It is akin to a fingerprint system for the digital world. We store the media files on a network of computers using IPFS, while blockchain maintains a record of ownership and changes that cannot be altered for any reason. To verify whether the media is authentic, we visually compare features with cosine similarity as facial recognition does with faces. Extra security measures were employed through BLAKE3 hashing as a one-fingerprint digital fingerprint at every checkpoint. We wanted to integrate a solution to protect digital media from tampering and fraudulent uses, yet as practical as possible in everyday life, generally, an upgrade or improved security system that offers the user ease of use.

Keywords Blockchain · IPFS · AES encryption · VGG16 · Digital media integrity · Web3.py · Metadata removal

1 Introduction

The digital revolution has fundamentally changed our interaction with visual media, presenting opportunities and challenges. Technology today opens up new avenues for creative expression, but there is also a growing concern with fact regarding the

R. Yegireddi (✉) · G. J. Sairam · V. Sandeep · M. Sirisha · B. L. Sivamani · M. Premsai
Department of CSE, Aditya Institute of Technology and Management, Tekkali, Andhra Pradesh, India
e-mail: rameshyegireddi@gmail.com

A. Kumar et al. (eds.), *Blockchain and Federated Learning Synergy for Privacy-Focused DeepFex Solutions*, Blockchain Technologies,
https://doi.org/10.1007/978-981-95-1394-9_12

authenticity of media via more complex manipulation options from simple edits to temporal adoptions of visual content via artificial intelligence that are now popularly termed deepfakes [1]. But with the availability of the platforms we have created, serious discussions are taking place regarding digital or media trust. The arena for verifying digital media is becoming more and more complex every day.

In a digital perfect storm of sorts, consumer editing applications are at the point, in many ways, where they rival professional software applications. Artificial intelligence systems can create or modify content that mimics on an incredibly realistic level, and social networks make modified content ubiquitous in a matter of minutes.

Lastly, most users do not have the level of technical expertise to detect these ways of altering digital content in unique ways, elevating their susceptibility to misleading content on digital platforms. In light of these increasing challenges, our research is proposing an enhanced framework for identifying and categorizing image modification that utilizes conventional means of analyzing images combined with VGG16 and built on blockchain authentication methods to increase the integrity of digital media, while protecting its proper applications in the domains of entertainment, art, and education.

Explanation of How the System Performs Media Authentication

The architecture develops three distinct components by adopting three separate technologies—VGG16's deep learning capabilities that can identify and extract prominent features in media, BLAKE3 speed and security for cryptographically hashing data in a fast manner, and a centralized InterPlanetary File System (IPFS)-blockchain approach for decentralized storage and verification of content. Our approach is made viable from three components that work together: a media processor that applies the initial analysis and security processes, a decentralized storage component that uses both an IPFS and blockchain storage method, and a content authentication engine that verifies the content using multiple checks and substantial work.

The framework uses a multilayer security model using a variety of technologies using advanced cryptographic techniques. The input media goes through AES encryption to provide confidentiality, and uses a blockchain to validate the media as authentic [2] (Fig. 1).

Fig. 1 Flow diagram of the blockchain-deep learning system

2 Literature Survey

Digital image encryption based on AES algorithm [3] by Zhang and Qunding proposes a secure image encryption system that takes a digital image, converts it to a 4 $\times$ 4 binary matrix, and executes the encryption using the AES encryption algorithm. With the help of histograms and key sensitivity tests, the method demonstrates strong security and shows complete sensitivity when converting a single-bit key. Overall, reconstruction accuracy was good, but with large images, there is room to improve processing speed, and toward improving the efficiency of real-time encryption.

Shamdasani et al. [4] proposed a hybrid approach that combines the capabilities of blockchain with the structure of IPFS to build a decentralized storage system that avoids the pitfalls of traditional cloud storage. The authors designed their framework to break data into separate units, perform a secure hash on each unit, and distribute all of the parts over the network infrastructure. The blockchain piece keeps the hash references of each distributed data to enable verification and authentication. In their study, the authors identified some challenges with implementation, which included limited capacity of the blockchain when considering large files, high costs of computation needed for mining, and using a modified version of the IPFS protocol to implement strong permissions provided by the owner for managing sensitive content.

Chang et al. [5] proposed an improved methodology for the detection of manipulated face images. Their work is based on modifications to the architecture of the VGG neural network. The authors developed advanced techniques for examining image noise and implemented robust data augmentations for improving detection performance. An important aspect of their implementation is that it contains a spatial richness model filter component that enhances the noise characteristics of manipulation artifacts in images. Their evaluations against the Celeb-DF benchmark showed performance that was superior to existing detection systems. Although the authors recognized challenges of keeping up with rapidly changing manipulation techniques and delivering consistent performance with a variety of image collections.

3 Methodology

3.1 Metadata Removal Process

The system undergoes a series of steps to improve security while maintaining the visual quality of the images. Initially, it transforms the binary file into a usable format, concentrating solely on the crucial pixel data as shown in Fig. 2. The system produces a refined version of the image, ensuring visual accuracy while eliminating embedded metadata like geolocation tags and device information [6].

This streamlined image undergoes final optimization, converting it to an efficient format such as PNG. The outcome is a visually identical image that lacks metadata

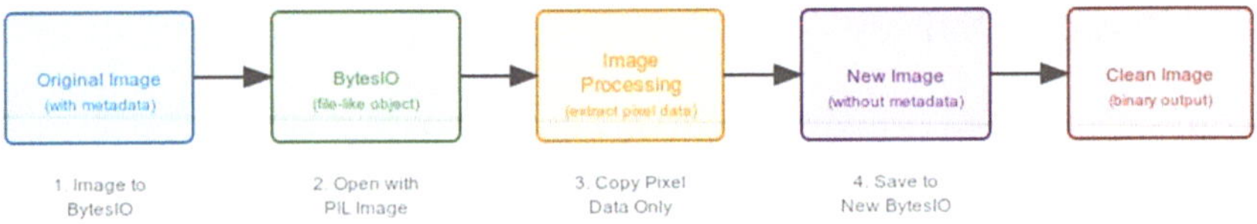

Fig. 2 Working of metadata removal process

but is specifically designed for blockchain storage and ensures user privacy without sacrificing image quality.

3.2 Media Upload Pipeline

3.2.1 Preprocessing

The upload pipeline utilizes a preprocessing approach for handling different image formats. The system processes two main image types:

Grayscale images encode only brightness values per pixel (0–255), making them ideal for applications where intensity matters more than color, like medical imaging. Converting grayscale to RGB [7] involves replicating the intensity value across all three color channels.

RGBA format combines standard RGB color channels with an alpha channel for transparency control, enabling advanced graphical effects and image overlays. The alpha values determine each pixel's opacity level.

All conversion processes maintain the original image's visual qualities while preparing it for future digital applications.

3.2.2 Feature Extraction

The system is using VGG16, which is a convolutional neural network, to extract features from images. There are three steps to the preprocessing pipeline:

1. Images are resized to 224 × 224 pixels so that they will be an input to VGG16. The resizing standardizes individual image sizes and allows the system to work with a uniform image size, regardless of the original image dimensions.
2. The resized images are converted to tensors—these are n-dimensional arrays that are supported to be handled by GPU-based computations and are subjected to extensive mathematical operations used in deep learning.
3. Tensor normalization scales pixel values according to the statistics of the ImageNet dataset:
4. Mean normalization values: [0.485, 0.456, 0.406]

5. Standard deviation values: [0.229, 0.224, 0.225].

This normalization allows input data distribution to become similar to VGG16's training data, leading to model improvement.

3.2.3 Feature Computation

After the preprocessing stages, the system processes the features in a two-step approach using the VGG 16 architecture [8]. Our method constructs strong feature representations while being computationally efficient.

The first stage convolves the normalized input using channel operation. For each of the color channels c in the preprocessed image I, the system calculates the normalized image representation:

$$I_c' = \frac{I_c - \mu_c}{\sigma_c}, \quad c \in \{R, G, B\} \tag{1}$$

where μ_c and σ_c represent the channel-specific mean and standard deviation values from the ImageNet statistics outlined in Sect. 3.2.2.

The second stage involves hierarchical feature extraction through VGG16's convolutional layers. At each layer l, the feature maps are computed using:

$$F_{\text{l}} = \max(0, W_{\text{l}} \times F_{\text{l}} - 1 + b_{\text{l}}) \tag{2}$$

where F_{l} represents the feature maps at layer l, W_{l} denotes the learned convolutional weights, b_{l} denotes bias, and the * symbol represents the convolution operation. The ReLU activation function, defined as $\max(0, \cdot)$, introduces nonlinearity, allowing the network to learn intricate visual patterns.

The final feature vector is derived by converting the output of the last convolutional layer into a one-dimensional representation:

$$v = \mathit{flatten}(FL) \in \mathbb{R}d \tag{3}$$

This process transforms the input image into a feature vector with d dimensions that encapsulates the hierarchical visual patterns learned by the network (Fig. 3).

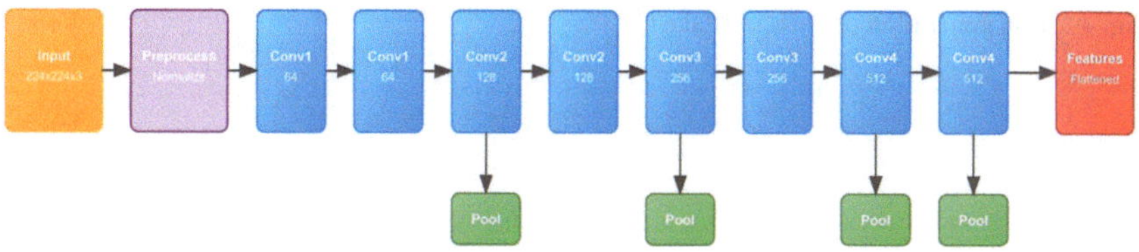

Fig. 3 Architecture of VGG16

3.2.4 Cryptographic Processing

The system uses cryptographic mechanisms as key components to ensure that the data created, received, or transmitted is secure in terms of its security, integrity, and confidentiality during its processes. In several ways, the system contains cryptographic aspects for secure treatment of image data through the processing pipeline.

1. Hash Generation

The system produces fixed-length digital fingerprints generated from the file or image binary data to provide unique identifiers to media assets. The implementation relies on the Blake3 algorithm after examining potential options because of its performance and security feature balances. The system converts the generated hash values into hex strings in a more readable and standardized form.

The Blake3 hash generation is mathematically defined as:

$$H(M) = h_2(IV, M, d) \tag{4}$$

where $H(M)$ is the final hash, M is input media bytes, IV is 8×32-bit initialization vector, h_2 is $2n$-bit compression function, and d is the desired output length in bits.

The compression function $\mathrm{h}_{2\mathrm{n}}$ in Blake3 operates using the following formula:

$$h_2(h, m, t, f) = G^7(h \| IV, m, t, f) \tag{5}$$

where the mixing function G applies seven rounds of operations, which include bitwise XOR, addition, right bit rotation, and left bit rotation (Fig. 4).

The algorithm's compression function accomplishes both mixing and scrambling properly, with hash responses that are resistant to both collision and preimage attacks. As a result, hash values for similar inputs, even with small differences in the input data, have hash values that are completely dissimilar. The hash process enables easy checking to see whether content has been altered or duplicate content exists.

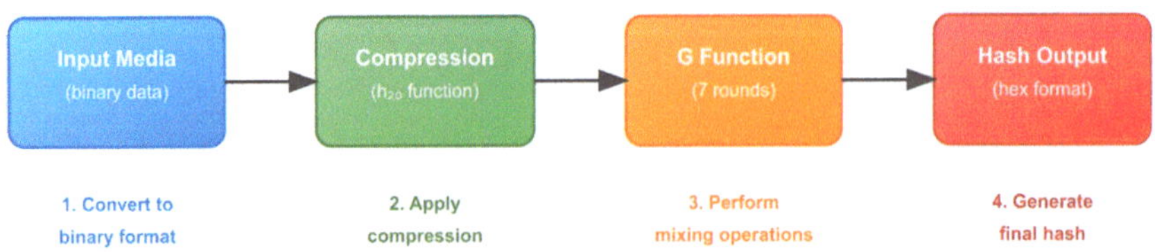

Fig. 4 Hash generation process in Blake3

4 AES Encryption

To protect media content, the system uses Advanced Encryption Standard (AES) to handle confidential data. AES is a symmetric-key block cipher that uses the same keys for encryption and decryption. The AES implementation depends on two elements of security: a cryptographically random key and an initialization vector (IV).

When AES is used in cipher block chaining (CBC) mode, the encryption occurs as follows:

For initialization:

$$C_0 = IV\,(\text{Initial Vector}) \tag{6}$$

For each block i of plaintext P:

$$C = E_k(P_i \oplus C_{i-1}) \tag{7}$$

In this expression, C_i represents the ith ciphertext block resulting from encryption, Pi indicates the ith block of plaintext data being processed, E_k refers to the advanced encryption standard algorithm implementation utilizing the designated encryption key, and IV designates the 16-byte initialization vector required for secure block cipher operation.

For PKCS#7 padding, $Pad(P) = P\|\{k, k, k, \ldots, k\}$, where $k = b - (|P| mod b)$, b is the block size (16 bytes), $|P|$ is the plaintext length, and || denotes concatenation.

The final encrypted output has the IV and then the content blocks that were encrypted. The IV needs to be stored with the ciphertext to enable decryption, as it is necessary for CBC mode.

Final ciphertext:

$$C = IV||C_1||C_2||\ldots||C \tag{8}$$

4.1 Blockchain Integration

4.1.1 Smart Contract Structure

The blockchain integration component of the system leverages Ethereum smart contracts to establish immutable storage capabilities for media authentication. The core smart contract defines an event that tracks three critical pieces of information: the encrypted media's Content Identifier (CID), the original media's cryptographic hash, and a CID for the associated features json file [9].

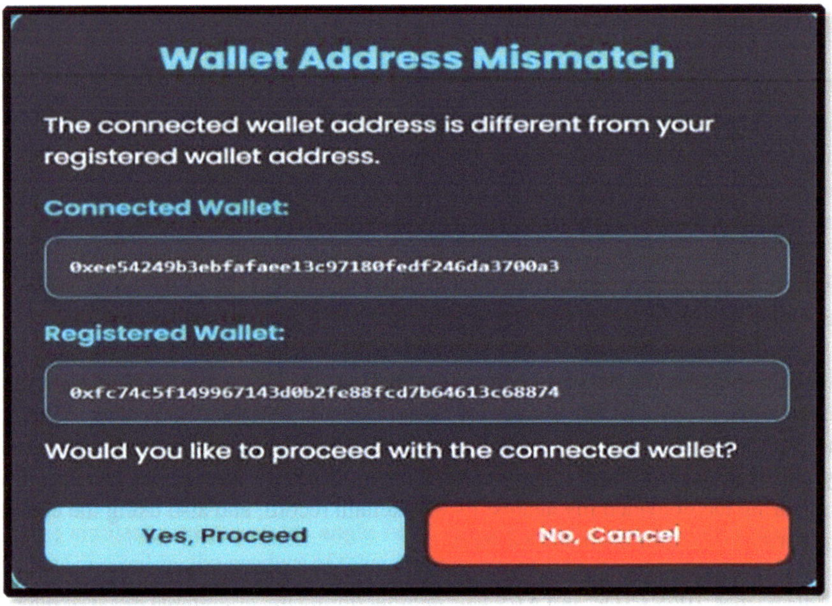

Fig. 6 Wallet address verification

Fig. 7 Dashboard page of the interface

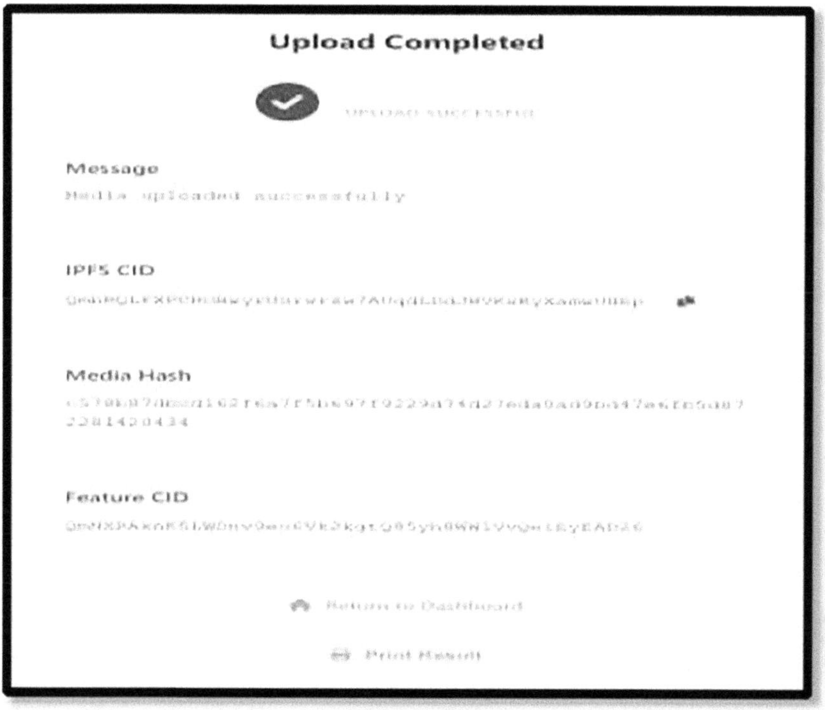

Fig. 8 Blockchain storage

Fig. 9 Verification page

6 Conclusion

Leveraging blockchain technology to combat deepfakes: a novel approach provides an innovative remedy to defend users' images from illegal use, offering a sound mechanism for authenticating media and blocking identity theft. As opposed to conventional techniques that cannot deal with detection accuracy and dependability, this method guarantees the secure authentication of images by associating them with unalterable blockchain records, enabling users to authenticate and monitor their media effortlessly. The system tackles fundamental issues like deepfake attacks and illicit uploads, ensuring greater security and transparency. Potential future studies and improvements can focus on enhancing scalability of the system, improving blockchain protocol for transaction processing at a quicker rate, and making it compatible with more platforms to ensure overall protection across the digital world. In addition, increasing the solution to incorporate higher-level authentication methods, including multifactor verification and also for video authentication, would increase its resilience.

References

1. Shaji George A, Hovan George AS (2023) Deepfakes: the evolution of hyper-realistic media manipulation, vol 1, no 2. Partners Universal Innovative Research Publication, pp 58–73. https://doi.org/10.5281/zenodo.10148558
2. Yazdinejad A, Parizi RM, Srivastava G, Dehghantanha A (2020) Making sense of blockchain for AI deepfakes technology. In: 2020 IEEE Globecom workshops (GC Wkshps), Taipei, Taiwan, pp 1–8. https://doi.org/10.1109/GCWkshps50303.2020.9367545
3. Zhang Q, Ding Q (2015) Digital image encryption based on advanced encryption standard (AES). In: 2015 Fifth international conference on instrumentation and measurement, computer, communication and control (IMCCC), Qinhuangdao, China, pp 1218–1221. https://doi.org/10.1109/IMCCC.2015.261
4. Shamdasani J, Deshmukh S, Wadile N, Sayyed M, Khandare S (2023) Decentralized file storage (interplanetary file system) using blockchain. Int J Eng Res Technol (IJERT) 12(03):252–256. https://doi.org/10.17577/IJERTV12IS030153
5. Chang X, Wu J, Yang T, Feng G (2020) DeepFake face image detection based on improved VGG convolutional neural network. In: 2020 39th Chinese control conference (CCC), Shenyang, China, pp 7252–7256. https://doi.org/10.23919/CCC50068.2020.9189596
6. Tayeb S et al (2018) Toward metadata removal to preserve privacy of social media users In: 2018 IEEE 8th annual computing and communication workshop and conference (CCWC), Las Vegas, NV, USA, pp 287–293. https://doi.org/10.1109/CCWC.2018.8301741
7. Jyothi G, Sushma CH, Veeresh DSS (2015) Luminance based conversion of gray scale image to RGB image. Int J Comput Sci Inf Technol Res 3(3):279–283. ISSN: 2348-120X
8. Jbara WA, Soud JH (2024) DeepFake detection based VGG-16 model. In: 2024 2nd international conference on cyber resilience (ICCR), Dubai, United Arab Emirates, pp 1–6. https://doi.org/10.1109/ICCR61006.2024.10533024
9. Khanum S, Mustafa K (2022) Sensitive data protection at blockchain endpoints and transactions. In: 2022 OPJU international technology conference on emerging technologies for sustainable development (OTCON), Raigarh, Chhattisgarh, India, pp 1–6. https://doi.org/10.1109/OTCON56053.2023.10113931
10. Alizadeh M, Andersson K, Schelén O (2020) Efficient decentralized data storage based on public blockchain and IPFS. In: 2020 IEEE Asia-Pacific conference on computer science and data engineering (CSDE), Gold Coast, Australia, pp 1–8. https://doi.org/10.1109/CSDE50874.2020.9411599
11. Nguyen HV, Bai L (2011) Cosine similarity metric learning for face verification. In: Kimmel R, Klette R, Sugimoto A (eds) Computer vision—ACCV 2010. Lecture notes in computer science, vol 6493. Springer, Berlin, Heidelberg. https://doi.org/10.1007/978-3-642-19309-5_55

MIX
Papier aus verantwortungsvollen Quellen
Paper from responsible sources
FSC® C105338

If you have any concerns about our products,
you can contact us on
ProductSafety@springernature.com

In case Publisher is established outside the EU,
the EU authorized representative is:
Springer Nature Customer Service Center GmbH
Europaplatz 3, 69115 Heidelberg, Germany

Printed by Libri Plureos GmbH
in Hamburg, Germany